Circuits and Systems for Biomedical Applications

EDITORS

Dr Hadi Heidari

University of Glasgow, UK

Dr Sara Ghoreishizadeh

University College London, UK

Tutorials in Circuits and Systems

For a list of other books in this series, visit www.riverpublishers.com

Series Editors

Peter (Yong) Lian

President IEEE
Circuits and Systems Society
York University, Canada

Franco Maloberti

Past President IEEE
Circuits and Systems Society
University of Pavia, Italy

River Publishers

Routledge
Taylor & Francis Group
LONDON AND NEW YORK

Published 2018 by River Publishers
River Publishers
Alsbjergvej 10, 9260 Gistrup, Denmark
www.riverpublishers.com

Distributed exclusively by Routledge
4 Park Square, Milton Park, Abingdon, Oxon OX14 4RN
605 Third Avenue, New York, NY 10158

First published in paperback 2024

Circuits and Systems for Biomedical Applications / by Dr Hadi Heidari, Dr Sara Ghoreishizadeh, Peter (Yong) Lian, Franco Maloberti.

Routledge is an imprint of the Taylor & Francis Group, an informa business

Publisher's Note
The publisher has gone to great lengths to ensure the quality of this reprint but points out that some imperfections in the original copies may be apparent.

While every effort is made to provide dependable information, the publisher, authors, and editors cannot be held responsible for any errors or omissions.

ISBN: 978-87-7022-053-8 (hbk)
ISBN: 978-87-7004-374-8 (pbk)
ISBN: 978-1-003-33754-6 (ebk)

DOI: 10.1201/9781003337546

Table of contents

Introduction

Circuit and system techniques envisioned to build wearable and implantable technologies over last several years. Large and ever-growing advances in developing and implementing such technologies exhibited the potential utility of this unique platform as the future of biomedical systems. Applications include lab-on-chip, lab-on-pill, Bio-inspired systems, bio-potential recoding, implantable neural stimulation and neuro-rehabilitation. The book through various chapters provides a deep insight into the technical problems that must be overcome to develop such complex circuits and systems. The rationale for selecting these paradigms is provided by a number of breakthroughs, which have been recently achieved in this field.

This book covers several advanced topics in the area of Wearable and ImplantableDevices, Analogue and Mixed-signal Circuits and Systems for applications in biomedical diagnostics, genomics, neural stimulation and rehabilitation. The fundamental aspects of these topics are discussed, and state-of-the-art developments are presented.

The book proceeds the 1st United Kingdom Circuits and Systems (UKCAS) Workshop (www. gla.uk/cas) and targets post-graduate students and young researchers as well as engineers working in industry willing to understand and connect circuit and system design with state-of-the-art and emerging medical applications. The field of biomedical electronics spanning from device technology to prosthetics limb is covered with emphasis on circuits and systems. The contents ensure a good balance between academia and industry, combined with a judicious selection of distinguished world-leading UK-based authors.

In **Chapter one** we will see how CMOS sensor technology has been extended into biotechnology and diagnostics such as for gene sequencing and biomarker detection on chip. We will introduce the emerging applications of CMOS sensors in metabolomics research and lab-on-chip. This is followed by **a detailed** description of a novel lab-on-a-pill endoscopic capsule in **Chapter two**. We present various stages in the development of the technology demonstrating the capsule is capable of capturing the contractile behavior of the Gastro-Intestinal. **Chapter three** introduces a novel nano-electronic technology, known as memristor, and presents few examples on how memristive technologies can be exploited in practical applications ranging from neuromorphic systems to charge-based computing and even enabling bioelectronics medicines.

Chapter four introduces analogue building blocks for neural-inspired circuits. In particular, a library of dynamically operating analogue functional blocks is described, based on principles, to emulate synaptic behavior, spike-dependent plasticity, synaptic fan-in, axonal delay and spike bursts. This is followed by introducing ideas for modelling of synaptic self-repair mediated by so-called astrocytes. **Chapter five** focuses on state-of-the-art low-power CMOS electronics for non-invasive and invasive bio-potential recoding (e.g. ambulatory EEG) from the central nervous system and the CMOS design methodologies related to such devices. In **chapter six** covers recent advances in scalable implantable neural stimulation systems using networked Application Specific Integrated Circuits (ASICs). Here we also discuss how such systems can meet the ever-growing demand for high-density neural interfacing and long-term reliability. **Chapter seven** summarizes CAS-related developments in the area of prosthetic limb and shows examples of success, in measurement of neuro-muscular activity, real-time processing and sensory feedback. The **last chapter** presents emerging CAS-enabled technology of genetically-enhanced brain implants for neuro-rehabilitation.

We hope that this book will serve as a useful resource to researchers and scientists in academia as well as industry in their effort to turn the new paradigm of emerging wearable and implantable technology into biomedical circuits and systems and to overcome bottlenecks in conventional silicon technology development.

We are grateful to all authors and speakers at the UKCAS workshop who have contributed their time and energy to make this book a reality. In their name,we also thank those people within their organizations who provided assistance to them. The compilation and editing of this book were, with great enthusiasm, supported by IEEE Circuits and Systems Society (CASS) and River Publishers.

Sara Ghoreishizadeh
Hadi Heidari
November 2018

CHAPTER 01

Integrated Circuits for Addressable Biosensing

David Cumming

University of Glasgow

Integrated circuit engineering has revolutionised computer and communication technology. However, silicon technology has also had a less expected role in the development of digital imaging, bringing CMOS to the fore as a sensing technology. In this talk I will present our work on extending CMOS sensor technology into biotechnology and diagnostics. In these applications, well-known sensors such as the photodiode can be exploited to integrate colorimetric assays on to an integrated circuit. It is also possible to integrate ion sensor technology what can be used for gene sequencing and biomarker detection on chip. Single photon avalanche diode technology has also been developed for use in diagnostics. The technology I will present has already made a major impact in genomics. More recent discoveries have shown that integrated circuit technology also has a role in metabolomics research and lab-in-a-pill technology.

1 MST@GU

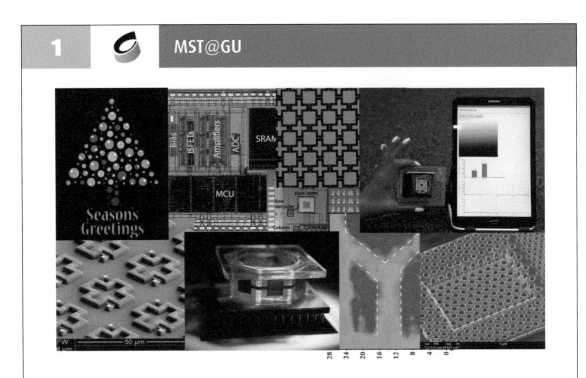

2 Overview

- **Microelectronics**
- **Sensors on CMOS**
- **Multiple similar measurements vs multiple different measurements**
  - Genomics vs metabolomics
- **Chip design – architecture for sensing**
- **Sensor types and post-processing**
- **Ionic**
- **Photonic**
- **Point of Care handheld and capsule**

3 Complementary Metal Oxide Semiconductor (CMOS)

- **Transistors!**
- **The core technology that underpins the microelectronics industry (hence everything else)**
  - Computers
  - Phones
  - Watches
  - Dishwashers
  - etc
  - Pretty much everything

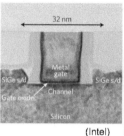

(Intel)

(Intel)

4 CMOS Photodiode (PD)

- A photodiode is a p-n junction photo sensitive detector
- A single pixel of a CMOS imager is an active pixel sensor (APS) that incorporates a PD and 3T for readout
- The amplitude of v_{out} is inversely proportional to the intensity of the detected light
- Transistor M3 controls connection to shared output line V_{out} so huge arrays of multiplexed sensors are made

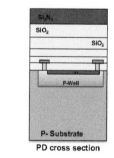

PD cross section

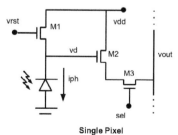

Single Pixel

5   Single Photon Avalanche Diode (SPAD)

- A SPAD is an avalanche photodiode reverse-biased above its breakdown voltage to operate in Geiger mode
- High electric field within the active region initiate an avalanche breakdown upon receiving a single photon
- The avalanche process creates a large current pulse that is detected by sensing circuitry
- Extends sensitivity beyond what can be achieved with a simple photodiode

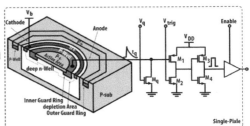

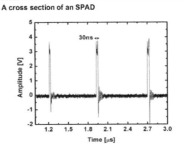

A cross section of an SPAD

6   Ion-sensing on CMOS

- Proton sensor (compare with photon sensor)

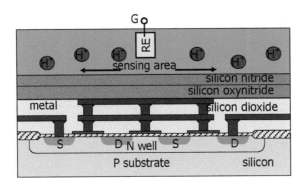

Ion Sensitive Field Effect Transistor

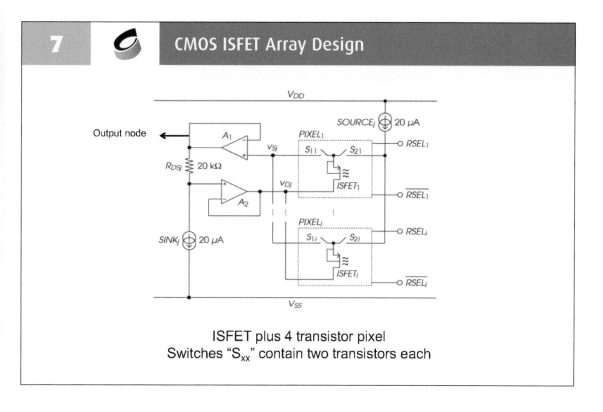

7 CMOS ISFET Array Design

ISFET plus 4 transistor pixel
Switches "S_{xx}" contain two transistors each

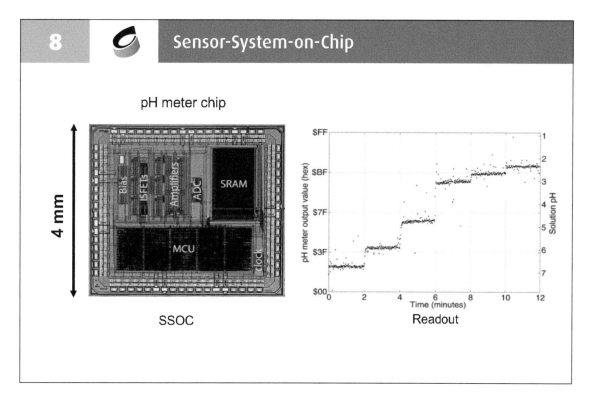

8 Sensor-System-on-Chip

pH meter chip

SSOC

Readout

MANY SIMILAR MEASUREMENTS

9 Imaging

10 Sequencing

- **Classic technology**

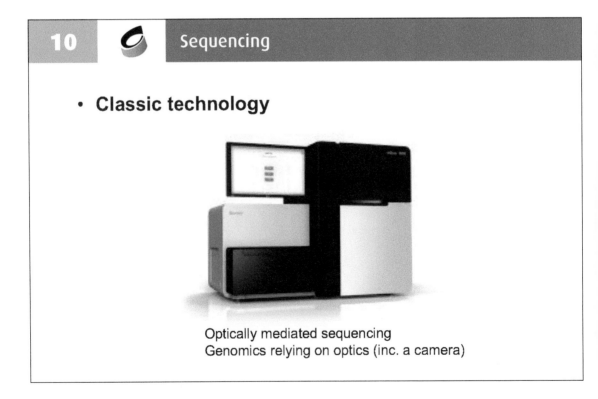

Optically mediated sequencing
Genomics relying on optics (inc. a camera)

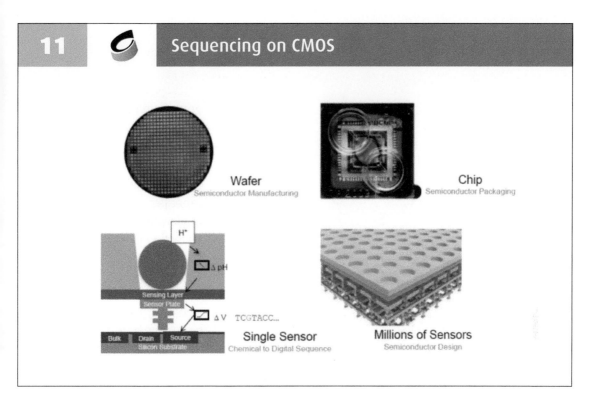

Sequencing on CMOS

Wafer
Semiconductor Manufacturing

Chip
Semiconductor Packaging

Single Sensor
Chemical to Digital Sequence

Millions of Sensors
Semiconductor Design

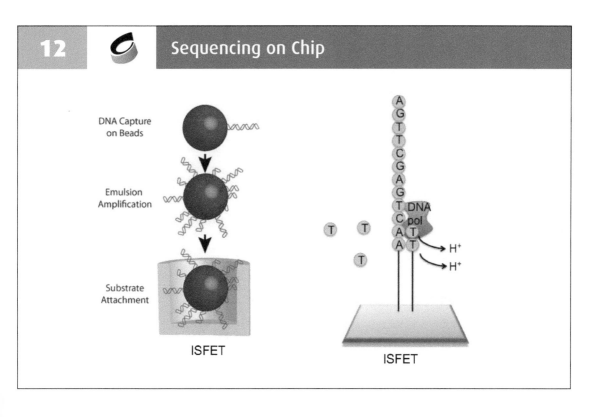

Sequencing on Chip

DNA Capture on Beads

Emulsion Amplification

Substrate Attachment

ISFET

ISFET

13

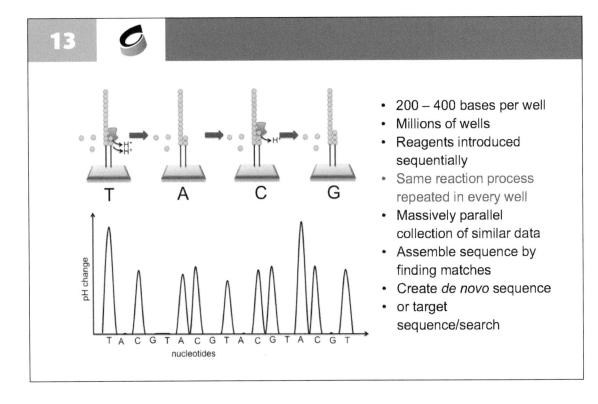

- 200 – 400 bases per well
- Millions of wells
- Reagents introduced sequentially
- Same reaction process repeated in every well
- Massively parallel collection of similar data
- Assemble sequence by finding matches
- Create *de novo* sequence
- or target sequence/search

14 The Personal Genome Machine

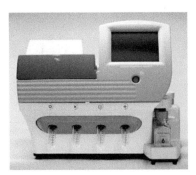

Former GU student,
Mark Milgrew

World's first optics-free sequencing system
Genomics relying on CMOS electronics
Once you have the chip, technically simple

DO THAT AGAIN FOR THE METABOLOME?

15

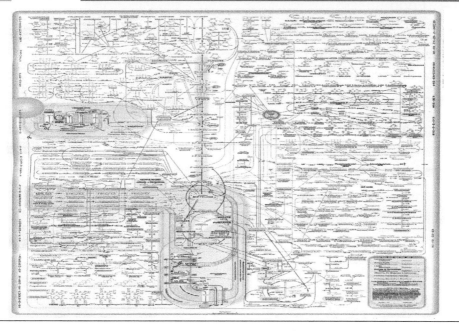

16 The metabolome vs the genome

- **The genome is chemically the same throughout**
  - Only four bases, A, T, C, G, repeated over and over
- **Gene expression environmentally controlled**
- **Hence the flourishing need for "polyomics"**
- **The metabolome is made up of several 1000 diverse molecules**
- **Differences may be dramatic**
  - Chemically different
- **Or they may be subtle**
  - e.g. structural
  - e.g. different monosaccharides

17 Examples of Metabolites

- **The small molecules of life, e.g.**
  - Food
  - Operational in metabolic (functional) pathways
- **Produce or breakdown food (e.g. respiration)**
  - Energy transport in blood
- **Make cellular components (e.g. proteins, RNA, DNA)**
  - Amino acids
- **Components of signaling pathways (e.g. controlling growth or cell differentiation)**
  - Tricarboxylic acid cycle (TCA) in mitochondria (respiration)
- **Response to infection (observed e.g. as inflammatory markers)**
  - C-reactive protein (CRP)

18 Direct measurement of metabolites

The diversity of the metabolome makes it hard to replicate the success of gene sequencing

Mass Spectroscopy

Nuclear Magnetic Resonance

- Equipment is large and costly

19   Diagnostic Panels

- **A stratified approach to diagnosis and treatment**
- **Many diagnostic "panels", e.g. stroke**

Measurand	Method of measurement	Relevance
Lactate (blood plasma)	Colorimetric assay	Increased in cerebral infarction. Linked to anaerobic glycolysis.
Pyruvate (blood plasma)		
BCAA (eg leucine, valine) (cerebral spinal fluid and plasma)	Laboratory spectrophotometric enzyme assay	Decreased in stroke.
Matrix metalloprotein-9 (MMP-9)	Enzyme-Linked Immunosorbent Assay (ELISA)	Increased after stroke.
Ubiquitin Fusion Degradation Protein 1 (UFD1) (plasma and serum)	ELISA	Increased in ischemic stroke.
Kynurenine (urine)	ELISA	Increased in stroke patients.

20   The Personal Metabolome Machine

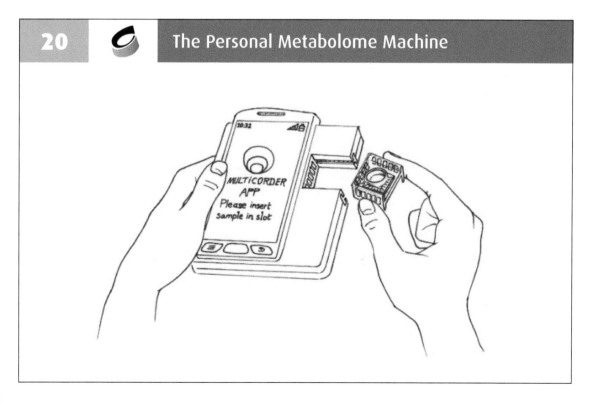

ENZYMES: AN ACCESS POINT FOR THE METABOLOME
Classic example is the blood glucose monitor used by diabetics

21 Enzymes

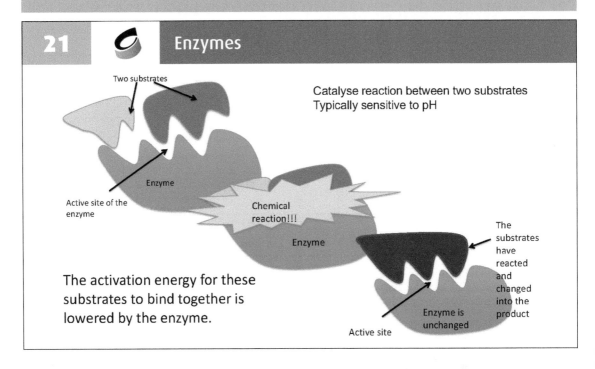

Two substrates

Catalyse reaction between two substrates
Typically sensitive to pH

Active site of the enzyme

Enzyme

Chemical reaction!!!

Enzyme

The substrates have reacted and changed into the product

The activation energy for these substrates to bind together is lowered by the enzyme.

Enzyme is unchanged

Active site

22 Example: The Glycolysis Cycle

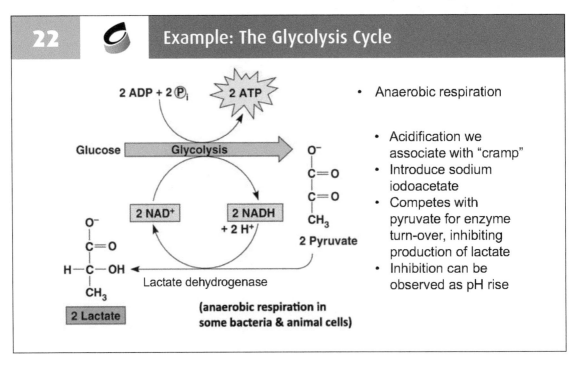

$2\ ADP + 2\ P_i$

$2\ ATP$

Glucose — Glycolysis

$2\ NAD^+$

$2\ NADH$
$+ 2\ H^+$

$2\ Pyruvate$

Lactate dehydrogenase

2 Lactate

(anaerobic respiration in some bacteria & animal cells)

- Anaerobic respiration

- Acidification we associate with "cramp"
- Introduce sodium iodoacetate
- Competes with pyruvate for enzyme turn-over, inhibiting production of lactate
- Inhibition can be observed as pH rise

23 Cell-based assay – metabolomic inhibition

- **16 x 16 ISFET array**

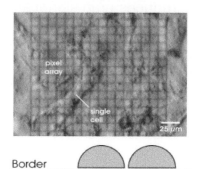

Border

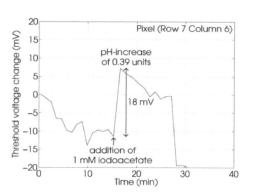

Glycolysis bio-assay
(leads to lactic acid build up in muscle)

- Cell culture grown over chip
- e.g. human fibroblasts
- Sensor pitch 11.5 µm

24 ISFET based assays

- Test substrate: glucose
- Glucose is processed by enzyme reaction
- Excess of protons is detected by pH sensing ISFETs

$$\text{D-Glucose} + \text{ATP} \xrightarrow{\text{Hexokinase}} \text{D-Glucose-6-P} + \text{ADP} + \text{H}^+$$

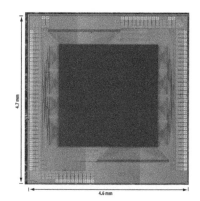

64k ISFET array

Test reaction vessel bonded to chip carrier

---

25 Results from hexokinase assay on ISFET

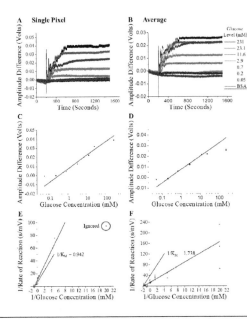

- Averaging across array reduces e^2 in signal and reduces discrepancies owing to mixing in vessel
- In excess of enzyme we measure initial velocity
- Michaelis-Menton constant = 0.57
- In reduced [enzyme] we measure [glucose] below physiological limit
- Note that hexokinase will turnover other sugars

---

26 Colorimetric Cholesterol Assay

$$Cholesteryl\ Ester + H_2O \xrightarrow{Cholesterol\ Esterase} Cholesterol + Fatty\ Acid$$

$$Cholesterol + O_2 \xrightarrow{Cholesterol\ Oxidase} 4 - Cholesten - 3 - one + H_2O_2$$

$$H_2O_2 + o - Dianisidine\ (reduced) \xrightarrow{Peroxidase} H_2O + o - Dianisidine\ (Oxidized)$$

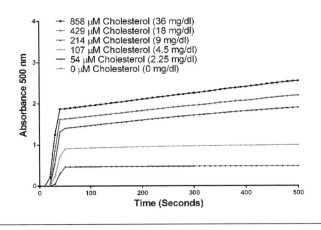

- O-Dianisidine is a colour change reagent
- Standard spectrophotometric assay
- We use this as a development tool for benchmarking our assays

27 Simplifying the optics

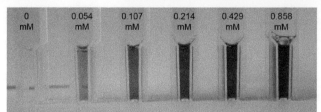

- **o-Dianisidine produces red/orange colour change**
- **Combine with sensitivity from PD**
- **Use green light from LED or monochromator**

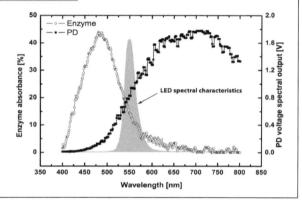

28 Colorimetric Sensing

- Limit of detection of 13 µM, 400 times lower than physiological range using photodiodes

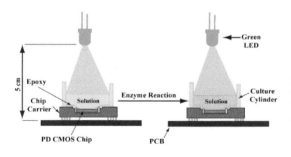

Simple illumination from LED c. 550 nm

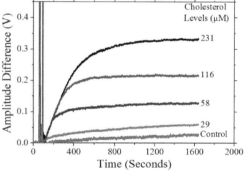

CIRCUITS AND SYSTEMS FOR BIOMEDICAL APPLICATIONS

---

29 **Chemiluminescence Sensing - demonstration**

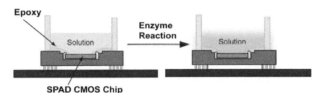

$$Luminol + H_2O_2 \xrightarrow{Peroxidase\ conjugated\ antibodies} Amino\ phthalat + N_2 + light$$

- Capability to detect varying hydrogen peroxide levels with peroxidase conjugated antibodies
- Immunoassay on chip without light source
- Uses our SPAD technology

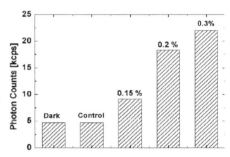

---

30 **Orthogonal sensing**

One-pot sensing

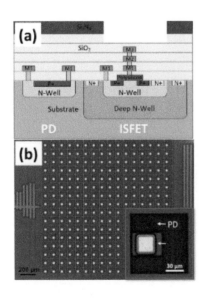

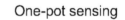

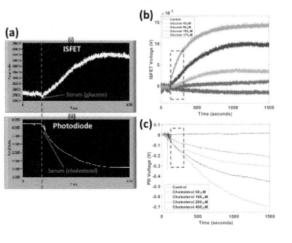

31 Tablet based data acquistion system

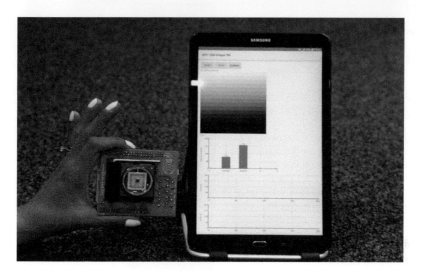

UNPUBLISHED

MINIATURISATION – SPADS IN CAPSULE ENDOSCOPY

32 The Diagnostic Pill

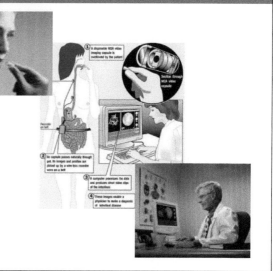

Focus on
bowel cancer screening (c.
16,000 deaths p.a. in UK)
Early diagnosis dramatically
improves outcome

33 Intestinal Imaging – upper GI tract

The Problem - Intestinal Imaging

- White light imaging suffers from high rate of false positives due reflective surfaces

- Autofluorescence imaging (AFI) can detect early signs of disease missed by white light

- AFI system (Olympus) requires high illumination (**up to 2mW**) endoscope-based, only covers oesophegus, large intestine

- Need for FI capsule that can take advantage of endogenous fluorophores (biological tissues) or can rely on specific exogenous markers with suitable fluorescent properties [1].

1. Transmitter
2. Controller
3. ASIC
4. Optical Filters
5. LED
6. Batteries

[1] M. B. Sturm, B. P. Joshi, S. Lu, C. Piraka, S. Khondee, B. J. Elmunzer, R. S. Kwon, D. G. Beer, H. D. Appelman, D. K. Turgeon, and T. D. Wang, "Targeted Imaging of Esophageal Neoplasia with a Fluorescently Labeled Peptide: First-in-Human Results," *Science Translational Medicine*, vol. 5, p. 184ra61, May 8, 2013 2013.

Given Imaging PillCam

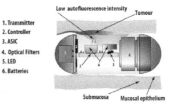

Low autofluorescence intensity — Tumour

Submucosa   Mucosal epithelium

Fluorescence Pill concept

34 32x32 SPAD imager capsule

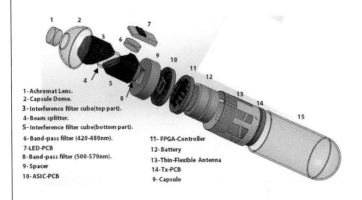

1 - Achromat Lens.
2 - Capsule Dome.
3 - Interference filter cube(top part).
4 - Beam splitter.
5 - Interference filter cube(bottom part).
6 - Band-pass filter (420-480nm).
7 - LED-PCB
8 - Band-pass filter (500-570nm).
9 - Spacer
10 - ASIC-PCB

11 - FPGA-Controller
12 - Battery
13 - Thin-Flexible Antenna
14 - Tx-PCB
9 - Capsule

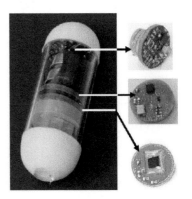

35   Imaging with 55 µW illumination

- **Autofluorescence imaging**
- **Fluorescence imaging**

Top row – FAD in water at different concentration.
Bottom row - 0.5mg of Haemoglobin is added to each one.
Measurement at 10ms SPAD Gating.

FITC at 100µM and at <1µM. 1.5 cm
distance and viewing angle > 50°.

AF intensity at 100ms exposure time.

45 µm-size fluorescence beads attached to a pig
tissue sample.

36   Conclusion

- CMOS has enabled several excellent sensor technologies to be made in a commercial process
- Access low cost technology for high value measurement
- Genomics accomplished
- Metabolomics more challenging
- Optical and direct chemical measurement possible
- Label and label- free methods
- The Personal Metabolome Machine

37 Microsystem Technology Group

- Visit www.gla.ac.uk/mst

- With thanks to all the members of the MST group, past and present and our academic and industrial collaborators

- Prof Mike Barrett, Glasgow University
- Prof Calum McNeill and Dr Neil Keegan, Newcastle University
- Prof Sandy Cochran, Glasgow University
- EPSRC Programme Grant "The Multicorder" and "SonoPill"

CHAPTER 02

Construction of an Endoscopic Capsule for the Diagnostics of Dysmotilities in the Gastro-intestinal Track

Marc Desmulliez

Heriot-Watt University

The presentation explains the work carried out in developing a novel endoscopic capsule, capable of capturing the contractile behaviour of the Gastro-Intestinal (GI) tract from multiple sites within the capsule. Such information potentially holds advanced diagnostic potential for timely detecting and diagnosing dysmotility disorders of the GI tract.

The research on the developed capsule was undertaken under the aegis of the EPSRC-funded Sonopill programme grant. Main contributors of this work were Mr Vasileios Mitrakos, Dr. Gerard Cummins and Dr.Sumanth Pavuluri. It was a joint collaboration with the University of Edinburgh under the supervision of Dr. Phil Hands. The vision of Sonopill was in developing an ingestible capsule, equipped with several sensor modalities for the diagnostics of pathologies with the Gastro-Intestinal Tract (GI) tract. The capsule would act thereby as a lab-in-a-capsule device for the continuous monitoring of the physiological parameters of the GI.

This presentation describes the sensing mechanism, the work carried out to develop a flexible pressure membrane, the reading system, in-vitro trials and animal models. The capsule effectively demonstrates movement of the bowel opening interesting prospects for capturing dysmotilities of the gut.

1 Outline

- Introduction
- Clinical motivation
- Technical motivation
- Performance of the pressure sensor
- PressureCap
- In vitro and in vivo trials
- Conclusions
- Outlook

2 Introduction

"Black-tube endoscopy," with the endoscope tethered by

an umbilical cord to a processor, will be viewed by future

generations with curiosity (Fleischer, 2010)

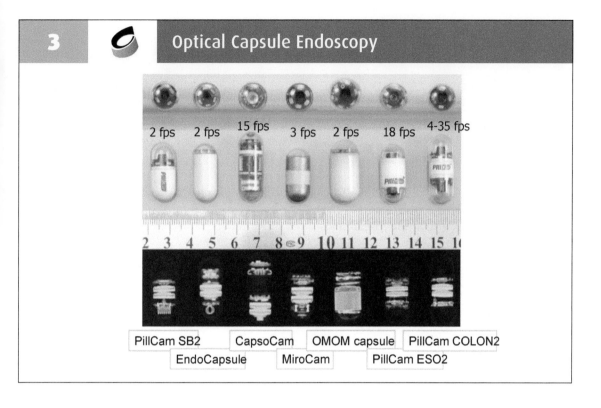

3 Optical Capsule Endoscopy

2 fps 2 fps 15 fps 3 fps 2 fps 18 fps 4-35 fps

PillCam SB2 CapsoCam OMOM capsule PillCam COLON2
EndoCapsule MiroCam PillCam ESO2

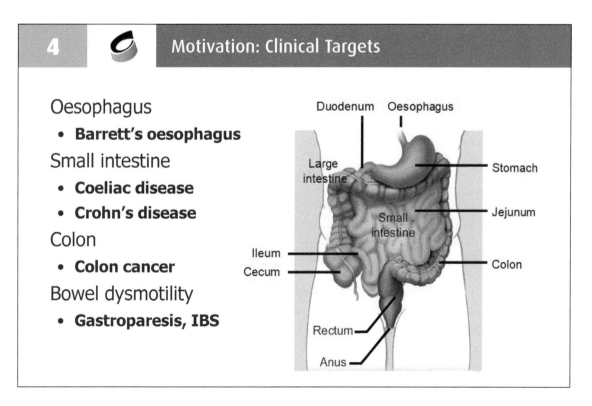

4 Motivation: Clinical Targets

Oesophagus
- **Barrett's oesophagus**

Small intestine
- **Coeliac disease**
- **Crohn's disease**

Colon
- **Colon cancer**

Bowel dysmotility
- **Gastroparesis, IBS**

Duodenum Oesophagus
Large intestine
Stomach
Jejunum
Small intestine
Ileum
Cecum
Colon
Rectum
Anus

5 Clinical motivation

- Monitoring of GI pressure useful for diagnosis of GI motility disorders (anorectal and oesophageal manometry)
  - *-eg. gastroparesis, irritable bowel syndrome, chronic idiopathic constipation, distinguishing visceral neuropathy from myopathy, gastrointestinal reflux disease.*

- Conventional endoscopic manometry catheters are:
  - *-invasive, uncomfortable and unable to reach most of the small intestine*

- PressureCap would provide an attractive alternative with additional functionalities:
  - *- capturing contractile pressure patterns and gradients*
    - *→ continuous non-invasive GI motility assessment*
  - *- supportive evidence of transition from one GI segment to another*
  - *- complementary role in measuring transit time*

6 Methodology

- Proof of concept capsules
  - Strain, temperature / humidity, single ultrasound elements, novel antennas, wireless power delivery

- Core capsules
  - Ultrasound array, fluorescence imaging, robotic positioning / localisation, multimodal diagnosis

  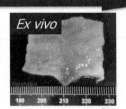 

- Therapeutic capsules
  - Ultrasound-mediated targeted drug delivery, microwave heating

- Exploratory capsules
  - Pressure, electro-physiology, infrared, gas sensing

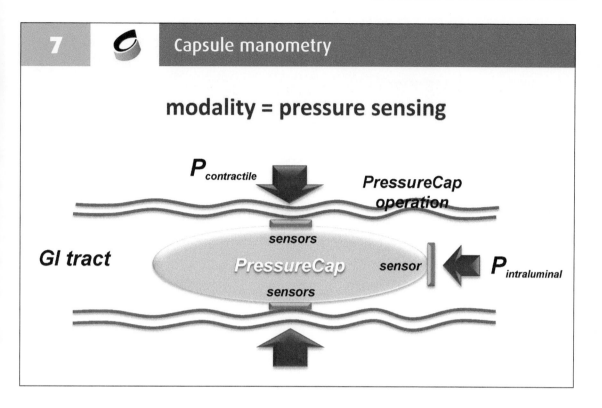

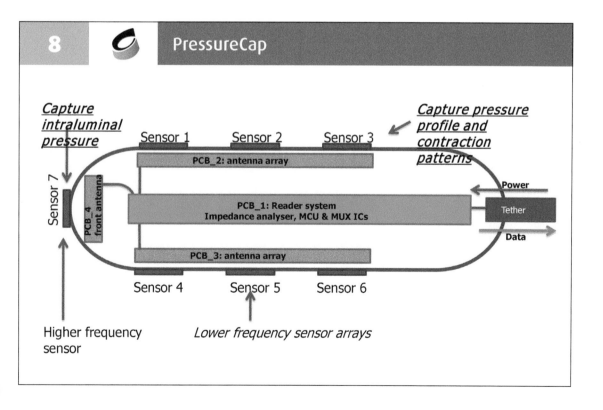

9 Sensor operation and modelling

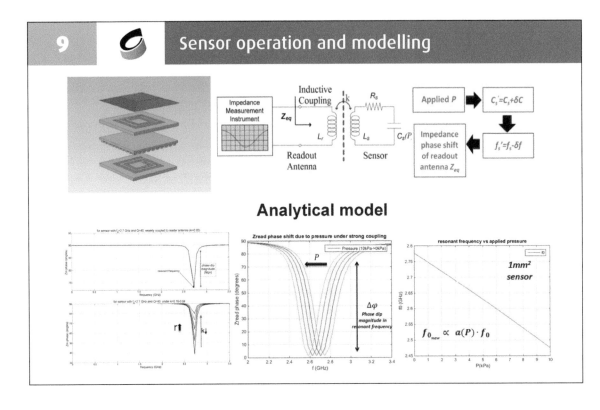

Analytical model

10 Pressure sensing technology

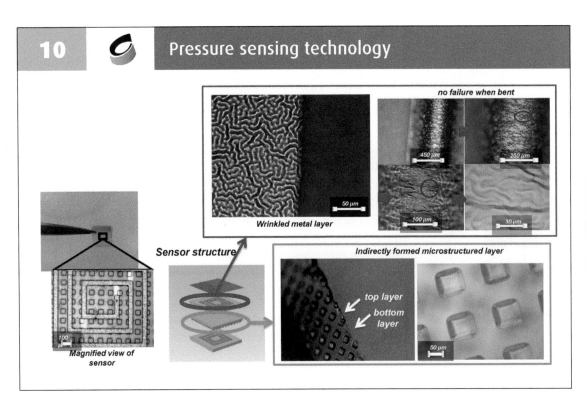

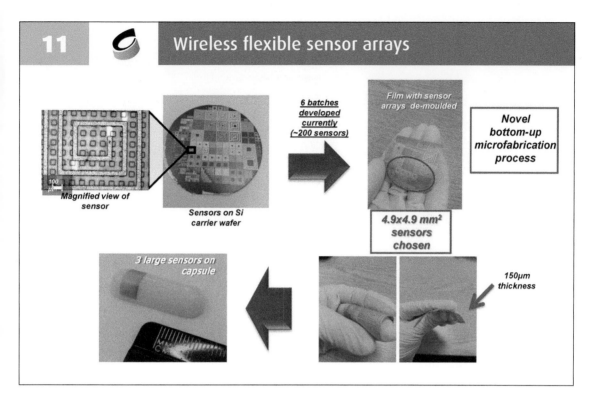

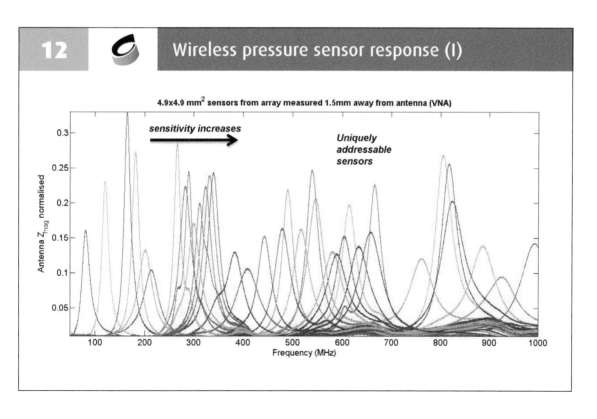

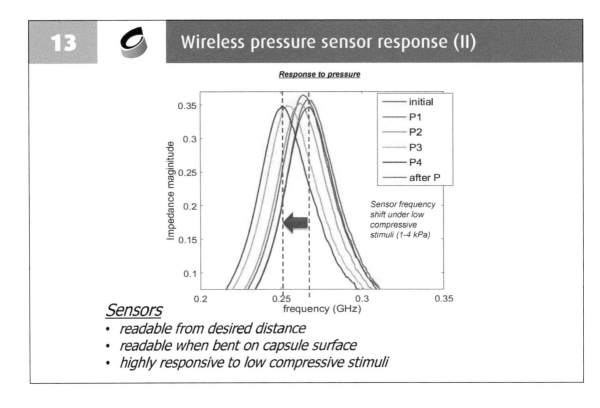

13 Wireless pressure sensor response (II)

Response to pressure

Sensor frequency shift under low compressive stimuli (1-4 kPa)

Sensors
- *readable from desired distance*
- *readable when bent on capsule surface*
- *highly responsive to low compressive stimuli*

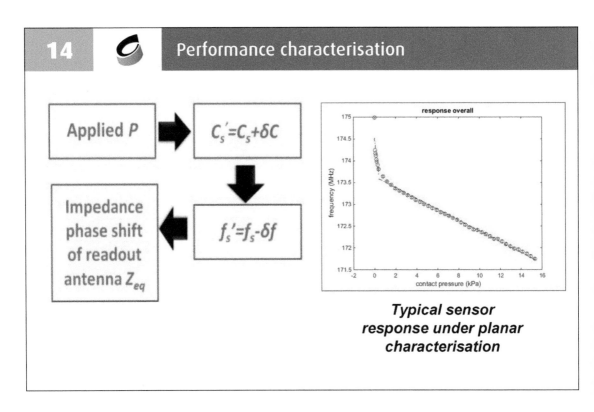

14 Performance characterisation

Applied *P* → $C_s' = C_s + \delta C$

↓

$f_s' = f_s - \delta f$

Impedance phase shift of readout antenna Z_{eq} ←

Typical sensor response under planar characterisation

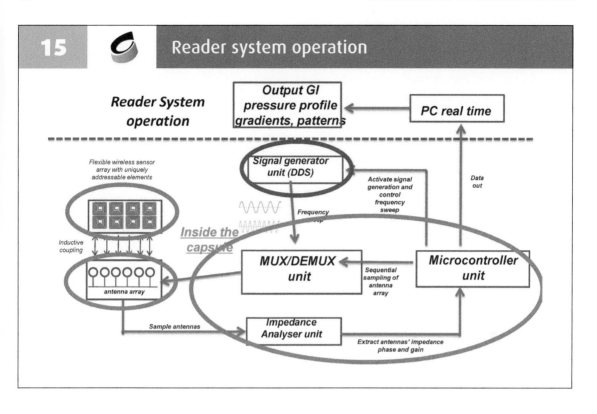

15 Reader system operation

Reader System operation

Output GI pressure profile gradients, patterns

PC real time

Flexible wireless sensor array with uniquely addressable elements

Signal generator unit (DDS)

Activate signal generation and control frequency sweep

Data out

Inside the capsule

Frequency sweep

Inductive coupling

antenna array

MUX/DEMUX unit

Sequential sampling of antenna array

Microcontroller unit

Sample antennas

Impedance Analyser unit

Extract antennas' impedance phase and gain

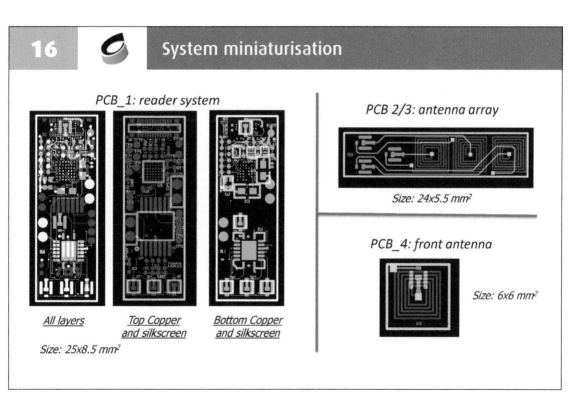

16 System miniaturisation

PCB_1: reader system

All layers

Size: 25x8.5 mm^2

Top Copper and silkscreen

Bottom Copper and silkscreen

PCB 2/3: antenna array

Size: 24x5.5 mm^2

PCB_4: front antenna

Size: 6x6 mm^2

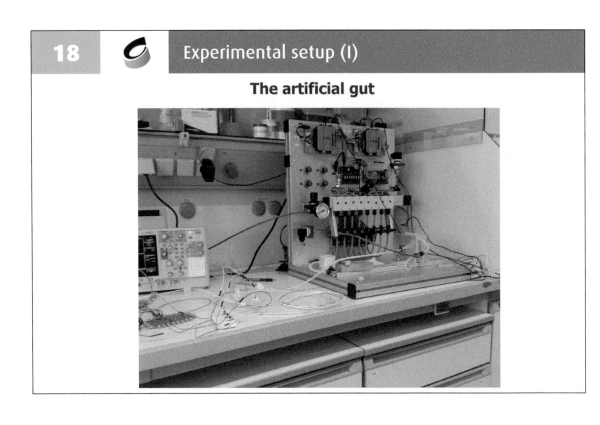

17   PressureCapassembly

Front antenna

Multi-antenna (x2)

Reader system component

Tether for data and power

Capsule with sensor array

System incorporated in capsule

Assembly of components

PressureCap top view

PressureCap front view

18   Experimental setup (I)

The artificial gut

19 Experimental setup (II)

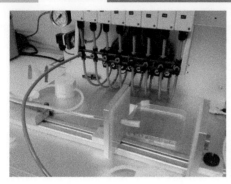

Capsule threaded through and fixed on position with the soft ecoflex stand

Ecoflex walls and capsule prior to incorporating the actuator gut unit

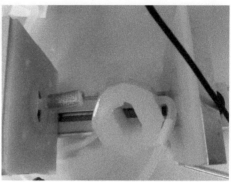

 20 Experimental setup (III)

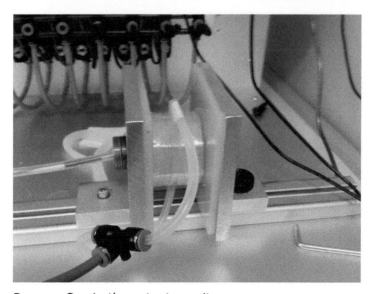

PressureCap in the actuator unit

21 The artificial gut

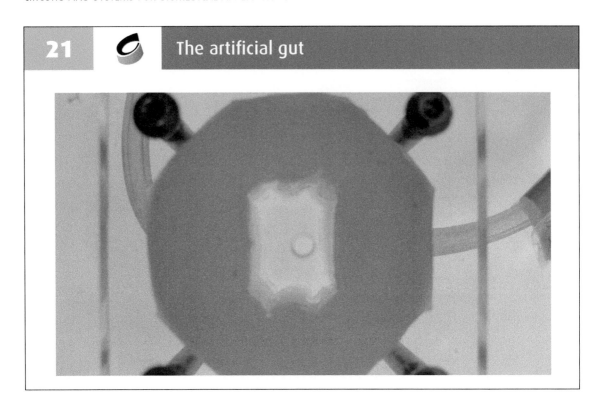

22 PressureCap *in vitro*

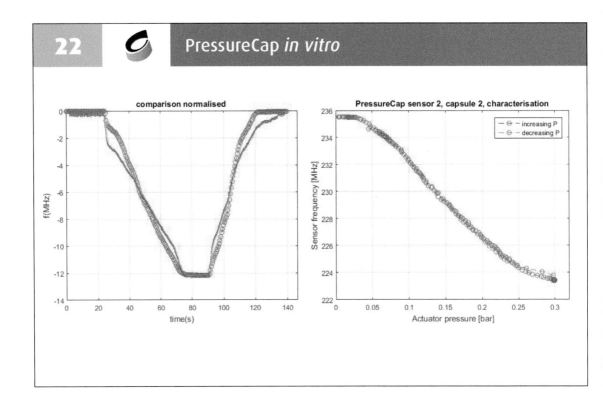

23   Peristalsis – Actuator measurements of phantom

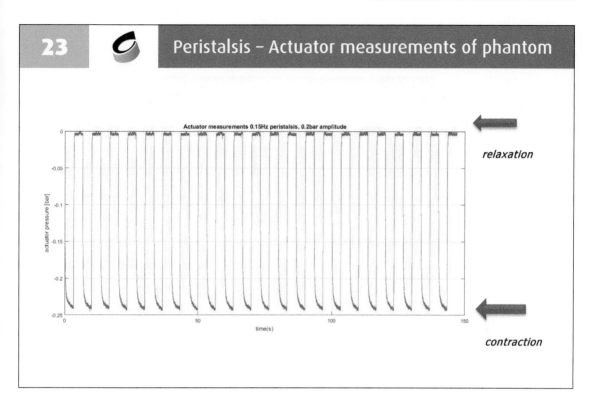

relaxation

contraction

24   Porcine trial (I)

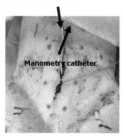

PressureCAP tether

Manometry catheter

Midline incision allowed the PressureCAP and catheter to be placed along same loop of bowel but 90cm apart

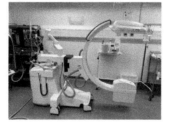

- C-Arm (Ziehm Vision) enabled conventional fluoroscopy and digital subtraction angiography (DSA) to be performed on subject using gastrografin as contrast agent.

- Protocol was to inject neostigmine, take measurements over a 20minute period and then start protocol again after a 5-10 minute break so capsule could be reset.

- Bowel motion clearly observed with fluoroscopy.

- Data from fluoroscopy, manometry catheter and PressureCAP for Pig 2 being analysed by Vasileios over next month or so

25 Porcine trial (II)

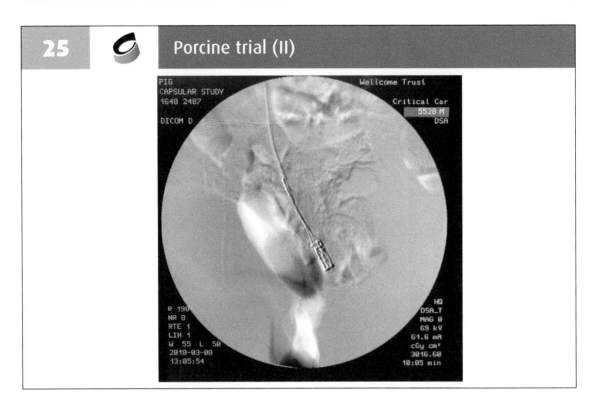

26 PressureCap *in vivo* trial

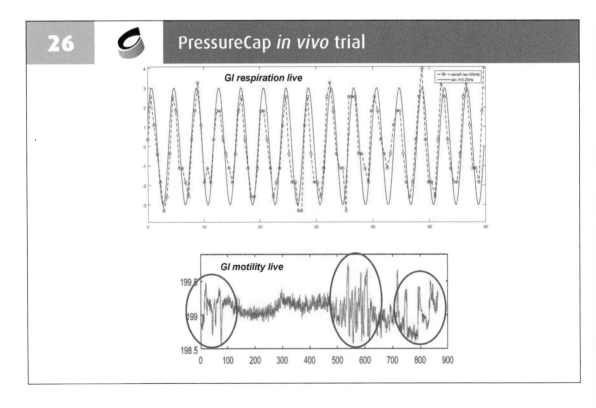

27 Conclusions

- Pressure sensor concept successfully validated

- More signal processing to be done to deconvolve respiration, heart beat and peristaltic motion

- More work on readout system to make it more performant and smaller

28 Outlook

Integration to wearable compression garments for continuous healthcare treatment efficacy monitoring *(currently)*

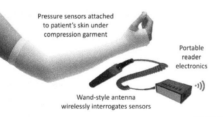

Pressure sensors attached to patient's skin under compression garment

Portable reader electronics

Wand-style antenna wirelessly interrogates sensors

Interpreted data displayed on portable device

Sensor #1 ✓
Sensor #2 ✓
Sensor #3 ✓

Fit OK

Intraocular Pressure (IOP) Monitoring

http://www.visionaware.org/info/your-eye-condition/glaucoma/patients-guide-to-living-with-glaucoma/125

Tactile sensing for robotic hand

https://www.vmt.nl/Nieuws/FTNON_
neemt_meerderheidsbelang_in_Lacqu
ey-150303133451

29 Acknowledgements

- Mr. Vasileios Mitrakos, HWU and UoEPhD

- Dr. Phil Hands, UoE

- Dr. Gerard Cummins, Dr Sumanth Pavuluri, HWU

- Dr. Ben Cox, Dundee University

- Prof. Eddie Clutton, Dryden Farms (LARIF)

- Ms. Ariane Tomas, ENSIRB (Bordeaux, France)

- Mr. Faulk Esser, Prof.Thomas Speck, University of Freiburg

- Ananth Technologies Ltd, Hyderabad, India

CHAPTER 03

Harnessing the Power of the Brain with Memory-resitors

Themis Prodromakis

University of Southampton

In the not so far future, electronic devices will be everywhere – embedded within our physical world and even in our bodies – empowering modern societies with unprecedented capabilities. Yet, the technological progress that brought us the mobile revolution is not any more sustainable for allowing us reaching this point. Up until now, the processing of data in electronics has relied on assemblies of vast numbers of transistors – microscopic switches that control the flow of electrical current by turning it on or off. Transistors have got smaller and smaller in order to meet the increasing demands of technology, but have nowadays reached their physical limit, with – for example – the processing chips that power smartphones containing an average of five billion transistors that are only a few atoms wide.

A novel nano-electronic technology, known as the memristor, proclaims to hold the key to a new era in electronics, being both smaller and simpler in form than transistors, low-energy, and with the ability to retain data by 'remembering' the amount of charge that has passed through them – akin to the behaviour of synaptic connections in the human brain. In his lecture Themis Prodromakis will present a few examples on how memristive technologies can be exploited in practical applications ranging from neuromorphic systems to charge-based computing and even enabling bioelectronics medicines.

1 Outline

- Modern electronics challenges

- Memristors: Technology, Tools & Infrastructure

- Applications: Examples – beyond memory

- What does the future look like?

MODERN ELECTRONICS CHALLENGES

2 The end of Moore's law?

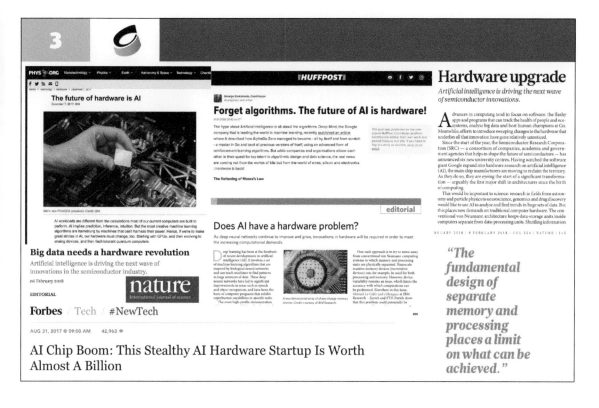

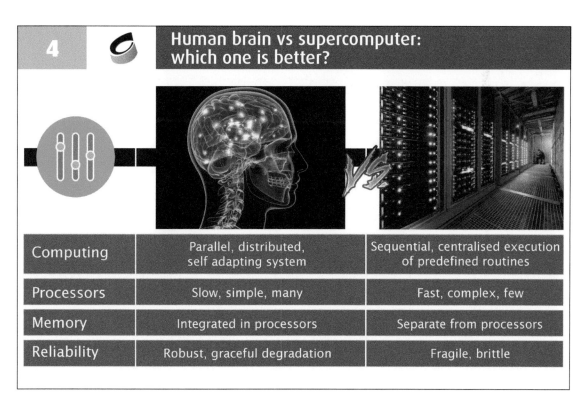

Computing	Parallel, distributed, self adapting system	Sequential, centralised execution of predefined routines
Processors	Slow, simple, many	Fast, complex, few
Memory	Integrated in processors	Separate from processors
Reliability	Robust, graceful degradation	Fragile, brittle

THE TECHNOLOGY

5 Memristor (Memory-resistor)

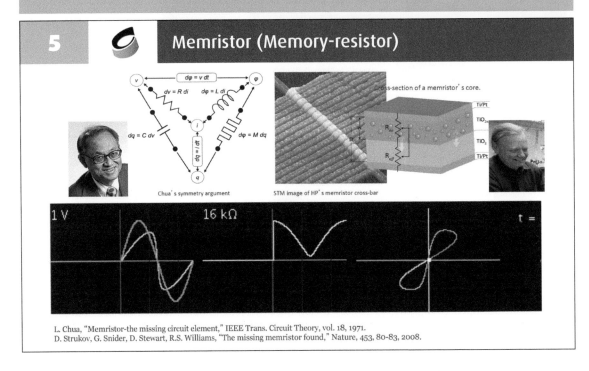

Chua's symmetry argument

STM image of HP's memristor cross-bar

Cross-section of a memristor's core.

L. Chua, "Memristor-the missing circuit element," IEEE Trans. Circuit Theory, vol. 18, 1971.
D. Strukov, G. Snider, D. Stewart, R.S. Williams, "The missing memristor found," Nature, 453, 80-83, 2008.

6 Memristors' hype cycle

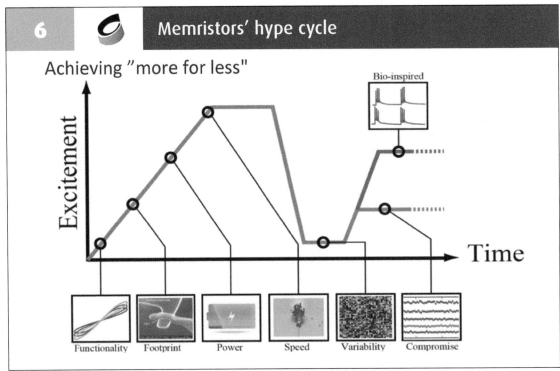

Achieving "more for less"

7   Memristors fabrication (I)

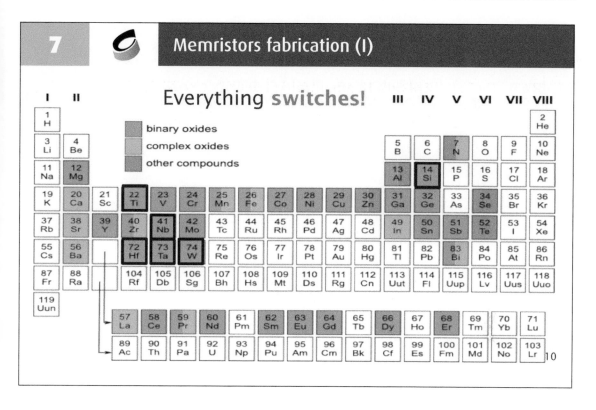

Everything **switches!**

- binary oxides
- complex oxides
- other compounds

8   Memristors fabrication (II)

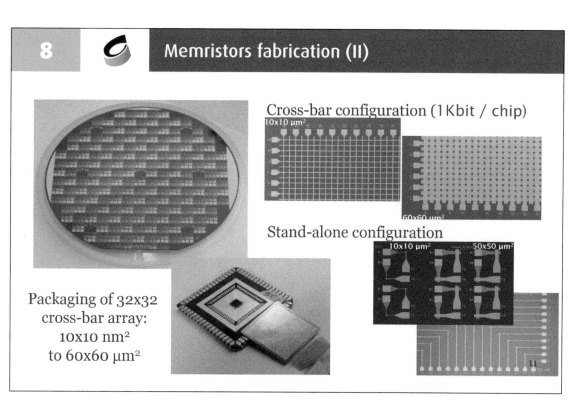

Cross-bar configuration (1 Kbit / chip)

$10 \times 10 \ \mu m^2$

$60 \times 60 \ \mu m^2$

Stand-alone configuration

$10 \times 10 \ \mu m^2$ $50 \times 50 \ \mu m^2$

Packaging of 32x32
cross-bar array:
$10 \times 10 \ nm^2$
to $60 \times 60 \ \mu m^2$

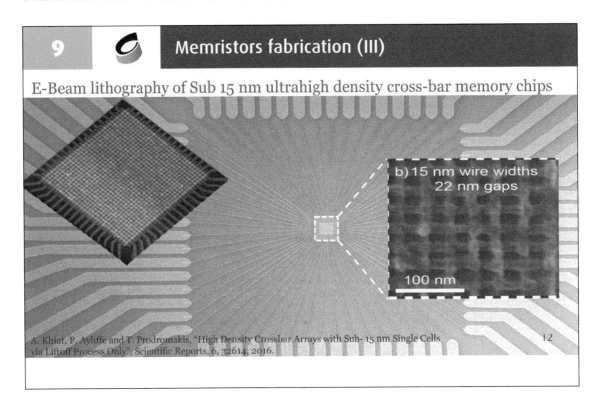

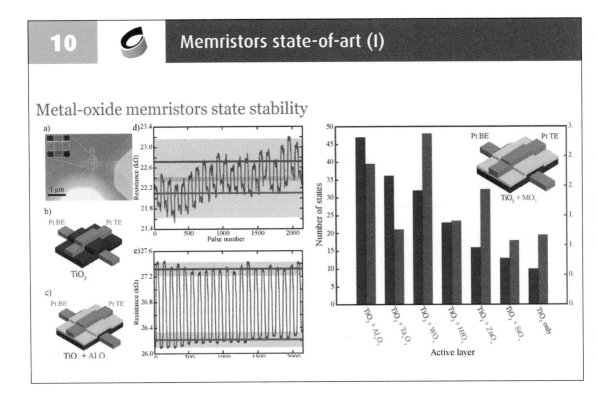

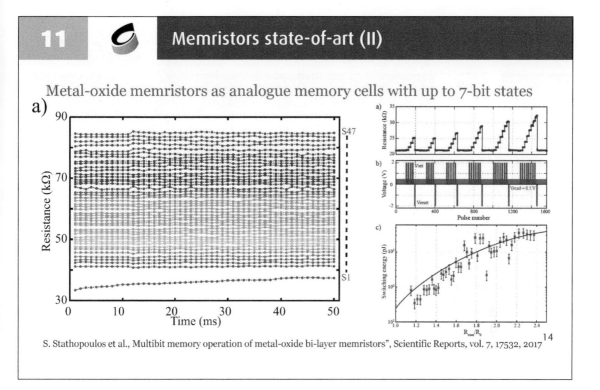

11   Memristors state-of-art (II)

Metal-oxide memristors as analogue memory cells with up to 7-bit states

S. Stathopoulos et al., Multibit memory operation of metal-oxide bi-layer memristors", Scientific Reports, vol. 7, 17532, 2017   14

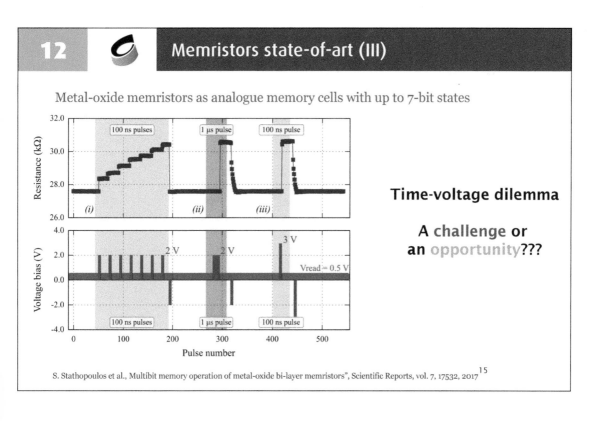

12   Memristors state-of-art (III)

Metal-oxide memristors as analogue memory cells with up to 7-bit states

Time-voltage dilemma

A challenge or an opportunity???

S. Stathopoulos et al., Multibit memory operation of metal-oxide bi-layer memristors", Scientific Reports, vol. 7, 17532, 2017   15

TOOLS & INFRASTRUCTURE

13

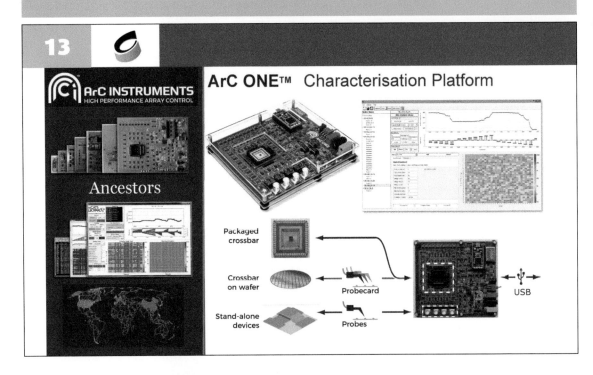

14 Array Control Instruments (I)

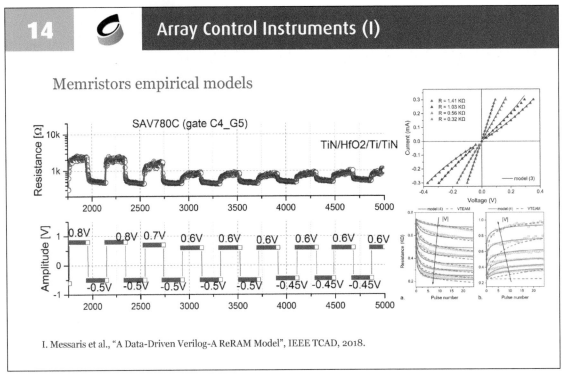

I. Messaris et al., "A Data-Driven Verilog-A ReRAM Model", IEEE TCAD, 2018.

15   Array Control Instruments (II)

Developing control instruments and modules for memristor arrays

L. Michalas, A. Khiat, S. Stathopoulos, and T. Prodromakis, "Conduction mechanisms at distinct resistive levels of Pt/TiO2-x/Pt memristrors", Appl Phys Lett, in press.

www.arc-instruments.co.uk

APPLICATION DEMONSTRATORS EXAMPLES – BEYOND MEMORY

Example #1

16   Emulating synapses with memristors

R. Berdan et al., "Emulating short-term synaptic dynamics with memristive devices", Scientific Reports, 6, 18639, 2016.
S. L. Wei et al., "Emulating long-term synaptic dynamics with memristive devices", arXiv, 2015.

17 Unsupervised Learning (I)

Unsupervised learning in probabilistic memristor neural network

Switching vs. resistive state relation at fixed voltage levels -> Exploit to encode conditional probabilities

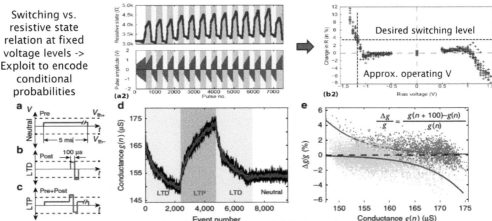

A. Serb, J. Bill, A. Khiat, R. Berdan, R. Legenstein and T. Prodromakis, "Probabilistic neural networks with multi-state metal-oxide memristive synapses", Nature Communications, 7, 12611, 2016.

18 Unsupervised Learning (II)

Unsupervised learning in probabilistic memristor neural network

State dependent STDP allows memristor to encode $p(PRE|POST = 1)$ in equilibrium resistive state.

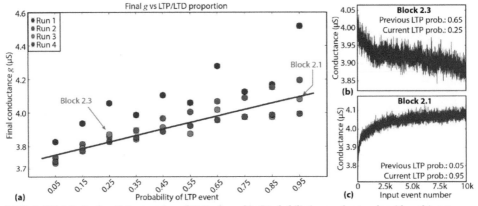

A. Serb, J. Bill, A. Khiat, R. Berdan, R. Legenstein and T. Prodromakis, "Probabilistic neural networks with multi-state metal-oxide memristive synapses", Nature Communications, 7, 12611, 2016.

19 — Unsupervised Learning (III)

Unsupervised learning in probabilistic memristor neural network

- Network shows capability of learning in unsupervised manner and <u>handles mistakes </u>rather well.

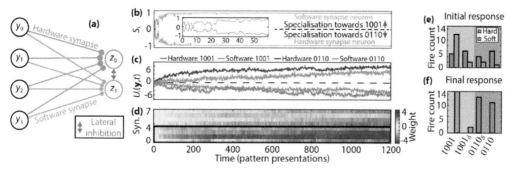

A. Serb, J. Bill, A. Khiat, R. Berdan, R. Legenstein and T. Prodromakis, "Probabilistic neural networks with multi-state metal-oxide memristive synapses", Nature Communications, 7, 12611, 2016.

20 — Unsupervised Learning (VI)

Unsupervised learning in probabilistic memristor neural network

- Whilst 'learn once' systems have their uses, ideally one wants something more flexible (e.g. if class centres drift over time).

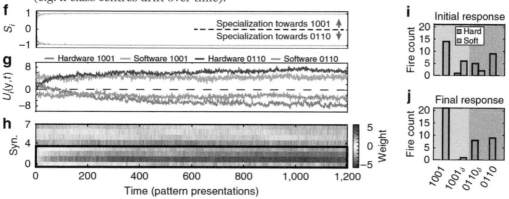

A. Serb, J. Bill, A. Khiat, R. Berdan, R. Legenstein and T. Prodromakis, "Probabilistic neural networks with multi-state metal-oxide memristive synapses", Nature Communications, 7, 12611, 2016.

Example #2

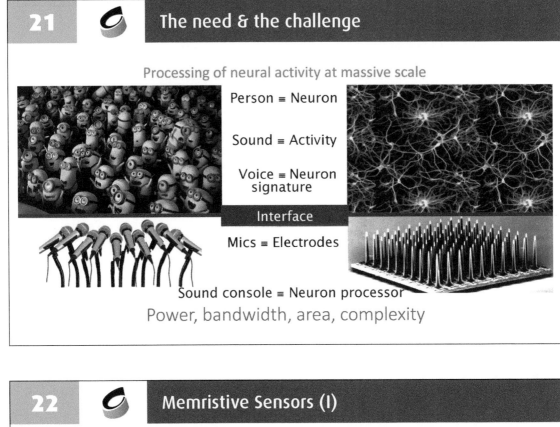

21 | The need & the challenge

Processing of neural activity at massive scale

Person ≡ Neuron

Sound ≡ Activity

Voice ≡ Neuron signature

Interface

Mics ≡ Electrodes

Sound console ≡ Neuron processor

Power, bandwidth, area, complexity

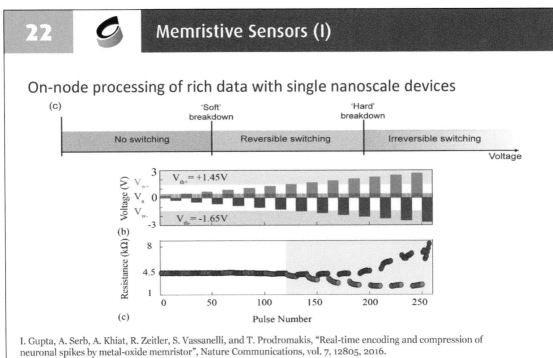

22 | Memristive Sensors (I)

On-node processing of rich data with single nanoscale devices

(c)

'Soft' breakdown   'Hard' breakdown

No switching   Reversible switching   Irreversible switching

Voltage

$V_{th+} = +1.45V$

$V_{th-} = -1.65V$

I. Gupta, A. Serb, A. Khiat, R. Zeitler, S. Vassanelli, and T. Prodromakis, "Real-time encoding and compression of neuronal spikes by metal-oxide memristor", Nature Communications, vol. 7, 12805, 2016.

23 **Memristive Sensors (II)**

On-node processing of rich data with single nanoscale devices

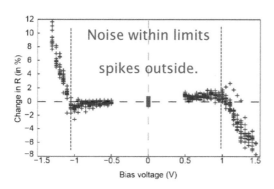

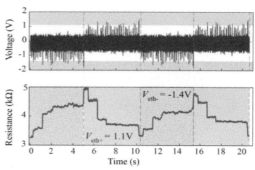

I. Gupta, A. Serb, A. Khiat, R. Zeitler, S. Vassanelli, and T. Prodromakis, "Real-time encoding and compression of neuronal spikes by metal-oxide memristor", Nature Communications, vol. 7, 12805, 2016.

24 **Monitoring large populations of neurons**

Towards building a HD analytical platform for cells

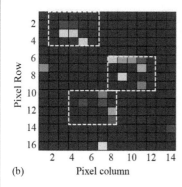
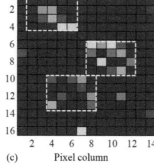
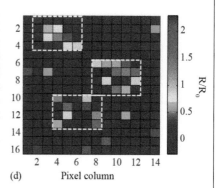

I. Gupta, A. Serb, A. Khiat, R. Zeitler, S. Vassanelli, and T. Prodromakis, "Real-time encoding and compression of neuronal spikes by metal-oxide memristor", Nature Communications, vol. 7, 12805, 2016.

25 Memristive Sensors (I)

Spike detection & sorting with single nanoscale devices

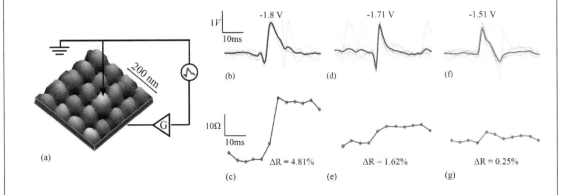

I. Gupta et al., "Spike sorting with metal-oxide memristors", Royal Society of Chemistry, Faraday Discussions, in press. arXiv:1707.08772

26 Memristive Sensors (II)

Spike sorting with single nanoscale devices

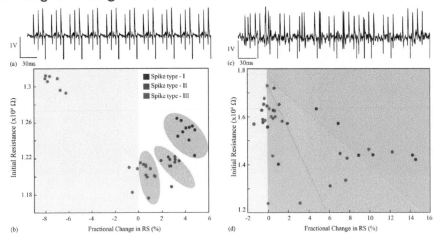

I. Gupta et al., "Spike sorting with metal-oxide memristors", Royal Society of Chemistry, Faraday Discussions, in press. arXiv:1707.08772

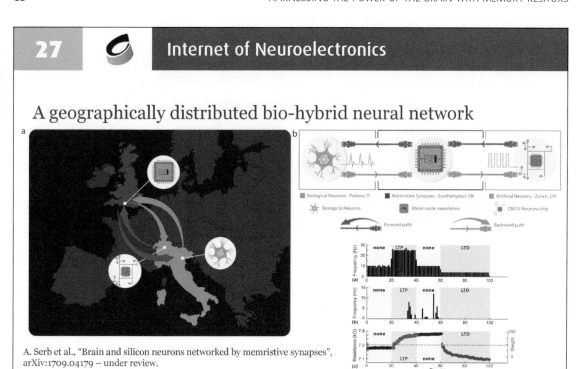

27   Internet of Neuroelectronics

A geographically distributed bio-hybrid neural network

A. Serb et al., "Brain and silicon neurons networked by memristive synapses", arXiv:1709.04179 – under review.

Example #3

28   Analogue – Digital

Our world is analogue!

Our electronics is mainly digital!

29 Charge-based computing (I)

Fusing Analogue and Digital Paradigms

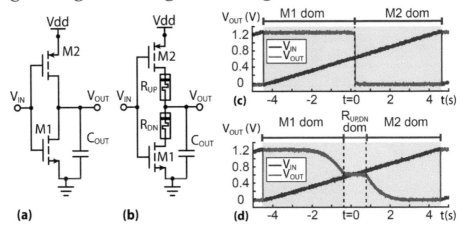

(a) **(b)** **(c)** **(d)**

A. Serb, A. Khiat and T. Prodromakis, "Seamlessly Fused Digital-Analogue Reconfigurable Computing using Memristors", Nature Comms, 9, 2170, 2018.

30 Charge-based computing (II)

Fusing Analogue and Digital Paradigms

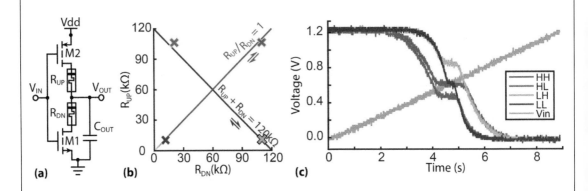

(a) **(b)** **(c)**

A. Serb, A. Khiat and T. Prodromakis, "Seamlessly Fused Digital-Analogue Reconfigurable Computing using Memristors", Nature Comms, 9, 2170, 2018.

31 — Charge-based computing (III)

Plane splitting in silico

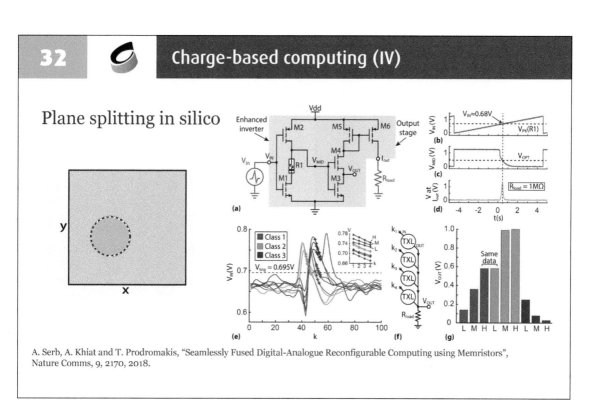

A. Serb, A. Khiat and T. Prodromakis, "Seamlessly Fused Digital-Analogue Reconfigurable Computing using Memristors", Nature Comms, 9, 2170, 2018.

32 — Charge-based computing (IV)

Plane splitting in silico

A. Serb, A. Khiat and T. Prodromakis, "Seamlessly Fused Digital-Analogue Reconfigurable Computing using Memristors", Nature Comms, 9, 2170, 2018.

WHAT DOES THE FUTURE LOOK LIKE?

33 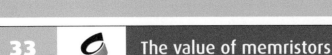 **The value of memristors**

Comparison to existing state-of-art technologies?

 2-3 orders of magnitude reduction in power

 x200 compression of information

 ~1 order of magnitude reduction in chip area

 Versatile deployment (adaptation, reconfigurable)

 Enable real-time giga-scale processing

 Maintain granularity of extracted information

34 **A pathway to keep your data private!**

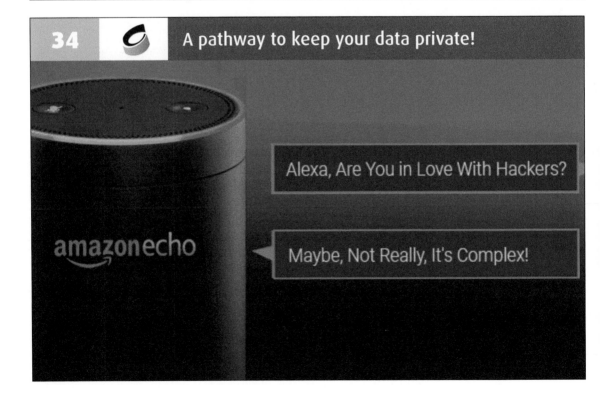

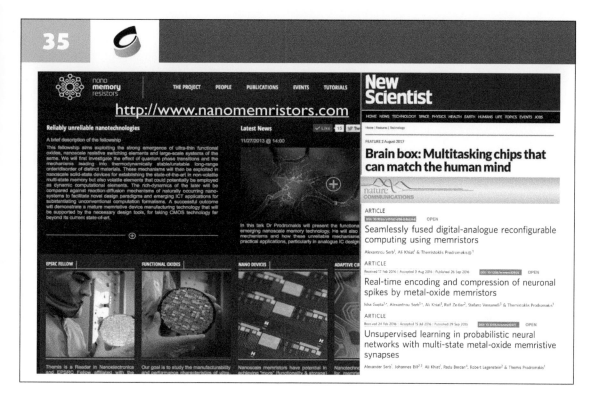

CHAPTER 04

Analogue building blocks for neural-inspired circuits

Steve Hall
University of Liverpool

Liam McDaid
Ulster University

Contact:

s.hall@liverpool.ac.uk
lj.mcdaid@ulster.ac.uk

Hardware implementation of neural networks is an increasingly popular area, both in research and industry. The main motivation is the amazing computational efficiency of the brain together with its self-repairing capability. The subject of artificial neural networks has a long history and much of the advance has been in software environments. Highly complex 'hidden layers' of synapses and neurons can be trained in so-called deep learning applications. Implementation in hardware however can bring significant advances in speed. Spiking neural networks (SNN) encode information in the timing of single spikes, and not just in their statistical firing rate [1]. Furthermore, recent neuroscience research has shown that SNNs mimic neuron behaviour on a level more closely related to biology and so have the propensity for powerful computational ability compared to classic artificial neural networks. Simulation platforms have been developed to allow investigation of the role played by spike-timing in the field of computational neuroscience [2]. Hardware approaches based on FPGAs, GPUs and ARM processor cores have been developed for hardware accelerated simulation which could offer such capability at the expense of large silicon area and low energy efficiency [3–5]. Good reviews of the current progress of hardware SNN development can be found in [6–8]. Different neuron models have been adopted varying from very detailed conductance-based approaches to simpler leaky integrate and fire versions. Conductance-based models emulate biophysical ion channels and hence are more faithful to biology. The integrate and fire models are somewhat less biologically realistic but require fewer transistors and can therefore be facilitated with compact layouts with low energy consumption. Thus, designers can balance accuracy with a higher number of neurons in the network, bringing the potential for better scalability. Each of the different implementations offers some trade-off between scalability, latency and biological realism. Finding an appropriate balance between these three elements represents one of the key challenges of hardware SNNs. In this chapter, the integrate and fire approach is explained. A library of dynamically operating analogue functional blocks is described, based on principles, to emulate synaptic behaviour, spike-dependent plasticity, synaptic fan-in, axonal delay and spike bursts. Some schemes for scaling are also described. Finally, ideas for modelling of synaptic self-repair mediated by so-called astrocytes are presented. References to associated publications can be found in the slide set.

[1] W Maass, 'Noisy spiking neurons with temporal coding have more computational power than sigmoidal neurons.' Adv. Neural Inf. Process. Syst. 9, 211–217 (1997),

[2] R Brette, 'Exact simulation of integrate-and-fire models with exponential currents.' Neural Comput. 19, 2604–2609 (2007)

[3] DB Thomas, W Luk, FPGA accelerated simulation of biologically plausible spiking neural networks, in: Proc. - IEEE Symp. F. Program. Cust. Comput. Mach. FCCM 2009, 45–52 (2009)

[4] AK Fidjeland, MP Shanahan, 'Accelerated simulation of spiking neural networks using GPUs.' 2010 IEEE World Congress (2010).

[5] SB Furber, F Galluppi, S Temple, LA Plana, 'The SpiNNaker project.' Proc. IEEE. 102, 652–665 (2014)

[6] G Indiveri, B Linares-Barranco, TJ Hamilton, A van Schaik, R Etienne-Cummings, T Delbruck, SC Liu, P Dudek, P Häfliger, S Renaud, J Schemmel, G Cauwenberghs, J Arthur, K Hynna, F Folowosele, S Saighi, T Serrano-Gotarredona, J Wijekoon, Y Wang, K Boahen, 'Neuromorphic silicon neuron circuits.' Front. Neurosci. (2011). doi:10.3389/fnins.2011.00073.

[7] J Misra, I Saha, 'Artificial neural networks in hardware: A survey of two decades of progress.' Neurocomputing. 74, 239–255 (2010)

[8] S Renaud, J Tomas, N Lewis, Y Bornat, A Daouzli, M Rudolph, A Destexhe, S Saighi, 'PAX: A mixed hardware/software simulation platform for spiking neural networks.' Neural Networks. 23, 905–916 (2010)

1 ## Some facts about the brain as a PC...

- The brain has ~100 billion neurons (10^{11}) – about 30μm large
  - Neuron Fan-in ~ $10^3 - 10^4$ (logic gates 2-4!)
  - complex dynamics - includes several time constants,
  - maintains a more complex internal state
  - output is a time-series of action potentials
    or 'spikes - no information in amplitude!
- Massively parallel in nature
  - Typical 10^{15} interconnections
  - Total computation rate of about 10^{16} complex operations /sec (cf 10 P-FLOPs)
- Millisecond time frame of 'events'
- Low level function: 'reasonably well understood'.
- High level function.....................???????

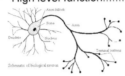

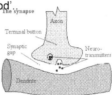

The brain is highly complex and no-one really understands how it works! It does incredible things, individual parts are seen to operate quite slowly but overall complex operations occur relatively quickly (thought to be due to parallelism) at incredibly low power: 10's W. 'Information' appears to be contained temporally, in spike chains from neurons but different trains have been observed to give the same action..

2 ## Some other brains

- A fly (1 grain of sugar a day to feed it!): 250 k neurons
- Honeybee (fantastic navigator!): 1 million neurons
- Rat (pretty smart animal): 55million neurons

- But how do the following work:
  - the arithmetic
  - Fault-tolerance
  - The parallelism (beat Moore's Law hands down)

> ### This is the inspiration!
> **But must find a simpler, scaleable, low power approach**

Many insects have 'distributed brains'; a relatively simple brain architecture interacting with a highly complex sensor network distributed throughout the body.

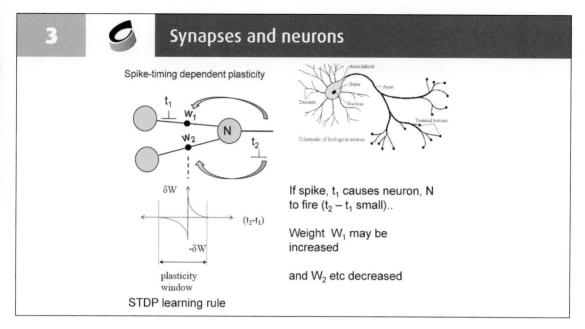

The basic units are synapses and neurons. Most of the 'intelligence' is associated with the synapse which is a complex electrochemical device. It is dealt with here in a rather simplistic manner. It receives spikes from a neuron and applies a weight to limit the amplitude, reflecting (to some extent) the 'importance' of the signal transmitted to the next neuron. There are many synapses feeding into a neuron as seen in the earlier slide. There is a competition as to which synapse causes the neuron to fire. The 'winners' are not necessarily the largest weights, the STDP learning rule also favours the last synapse(s) which cause the neuron to fire (emit a spike). This 'plasticity' is indicated in the learning rule shown on the slide.

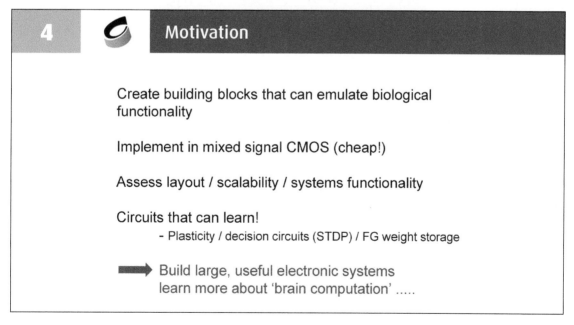

The talk is based on the results of an EPSRC funded project and this slide indicates our aims and objectives, namely to explore how close we could get to biology with the limitations of a charge-based technology.

5 Circuit Challenges

- Store and update weights
- Detect timing ($t_2 - t_1$)
- Axonal delay
- Low power operation
- Scale to VLSI
- Learn!

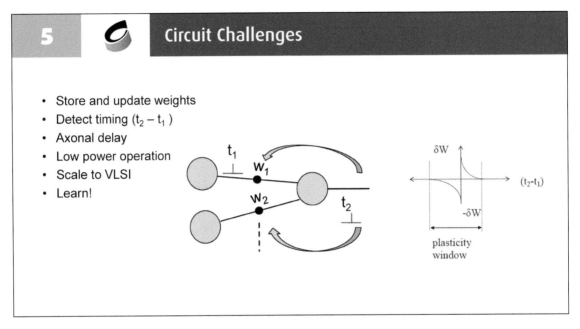

Each of these 'challenges' is possible to implement in Si circuitry/devices and in fact, most have been done. The real challenges are the scaleability,

6 Dynamic synapse

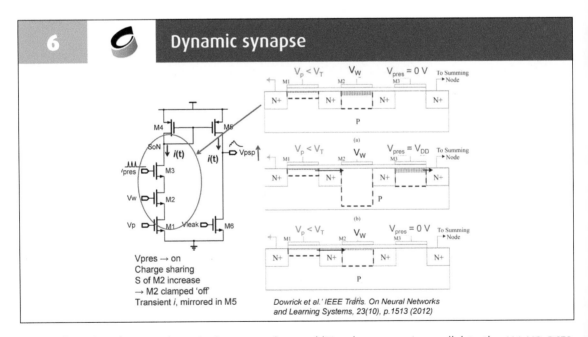

Vpres → on
Charge sharing
S of M2 increase
→ M2 clamped 'off'
Transient i, mirrored in M5

Dowrick et al.' IEEE Trans. On Neural Networks and Learning Systems, 23(10), p.1513 (2012)

The cell is referred to as a dynamic charge transfer synapse (DCTS). A single DCTS with summing node is shown, where transistors M1-M3 constitute the synapse and M4-M6 implement a summing function. The input connectivity can be increased by including additional synapses in parallel to the M1-M3 DCTS. While the synapse bears some similarity to existing implementations there are a number of important differences and associated advantages. Firstly, fewer MOSFETs are required to implement synaptic

6 **Dynamic synapse**

plasticity and depression than in other approaches, and the range of achievable recovery times for depression are greater. This significantly increases the packing density, which is vital when considering the construction of large scale networks. In addition, the circuit presented here operates in a fashion more akin to that of a charge couple device (CCD) than a regular MOSFET configuration, and can be deemed to be closer to the underlying biological principles.

The gate voltage of M2, V_W, sets the weight of the synapse, while the gate voltage of M1, V_P, controls the depressing effect and therefore determines the amplitude of the resulting PSP. The V_{PRES} terminal represents a spiking input from a pre-synaptic neuron. Biologically, a synapse is classed as depressing if it has a Paired Pulse Ratio (PPR) < 1. A neuron connected to a depressing synapse will produce a response which is depressing in nature. A PPR > 1 is representative of a facilitating synapse and will produce a facilitating PSP when connected to a neuron.

7 **How it works**

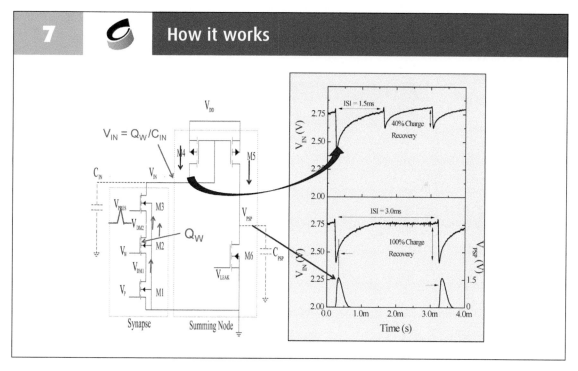

In a resting state, where V_{PRES} = 0V and voltages are present at V_W and V_P, the voltage at V_{IN} will sit at a value slightly less than V_{DD}, as there is a need for M4 to supply the small leakage current in the MOSFET chain. Drain voltages, V_{DM1} and V_{DM2} will be ~0V, to be consistent with the very low current in the MOSFET chain. Provided that V_W and V_P > 0V, the arrival of a voltage pulse, V_{PRES} of magnitude V_{DD}, and width ΔT, will initiate the flow of charge through the MOSFET chain. The presence of sub-threshold biased MOSFET, M1 however, will cause the MOSFETs

M2 and M3 to operate in a charge transfer mode; that is, the weighted charge slug will be effectively dumped on the node of V_{IN} where charge sharing occurs The voltage on the source of M_2 will jump up in value, clamping that MOSFET off; the effective substrate bias contributing to that condition. At the same time, the drain-source voltage of Mw will go close to zero. Once V_{PRES} goes to zero, these latter voltages will relax back to their starting conditions as the charge slug required under M2 is gradually provided by sub-threshold current supplied by M1.

1 How it works

It is also possible for the dynamic synapse device to be operated in an inhibitory fashion simply by connecting its output to the V_{PSP}, rather than the V_{IN}, node; where the arrival of a synaptic input will cause charge to be removed from, rather than added to, the node. Multiple synaptic inputs will be summed to produce a PSP, the shape of which is dependent on the values of the control voltages and interspike interval (ISI). If the amount of charge provided by additional synaptic inputs is greater than the amount of charge which leaks away during the ISI, then the PSP will continually increase. If the leakage charge is greater than the additional synaptic charge, then the PSP will reach a peak magnitude on the first spike, which will reduce for each subsequent spike. In general, if the applied ISI $\leq \tau_S$, then for a given value of V_{LEAK}, $V_P > V_{LEAK}$ will produce the former case and $V_P < V_{LEAK}$ will produce the latter.

8 Post-synaptic potentials

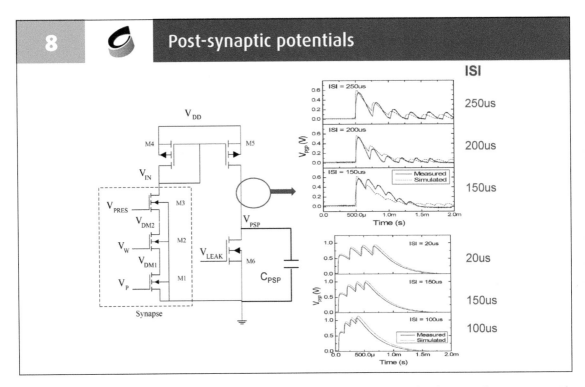

Examples of post-synaptic potential outputs from the DCTS. Experimental results show good agreement with simulations in the Cadence environment.

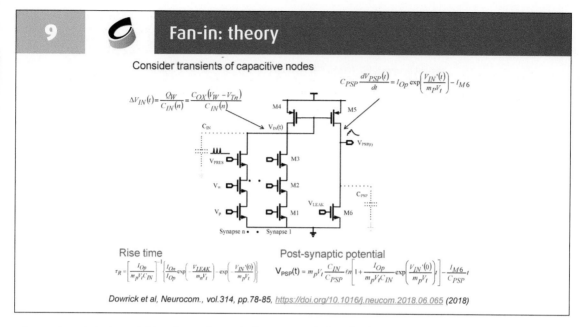

9 · Fan-in: theory

Consider transients of capacitive nodes

$$C_{PSP}\frac{dV_{PSP}(t)}{dt} = I_{Op}\exp\left(\frac{V_{IN}'(t)}{m_pV_t}\right) - I_{M6}$$

$$\Delta V_{IN}(t) = \frac{Q_W}{C_{IN}(n)} = \frac{C_{OX}(V_W - V_{Tn})}{C_{IN}(n)}$$

Rise time

$$\tau_R = \left[\frac{I_{Op}}{m_pV_tC_{IN}}\right]^{-1}\left\{\frac{I_{On}}{I_{Op}}\exp\left(\frac{V_{LEAK}}{m_nV_t}\right) - \exp\left(\frac{V_{IN}'(0)}{m_pV_t}\right)\right\}$$

Post-synaptic potential

$$V_{PSP}(t) = m_pV_t\frac{C_{IN}}{C_{PSP}}\ell n\left[1 + \frac{I_{Op}}{m_pV_tC_{IN}}\exp\left(\frac{V_{IN}'(0)}{m_pV_t}\right)t\right] - \frac{I_{M6}}{C_{PSP}}t$$

Dowrick et al, Neurocom., vol.314, pp.78-85, https://doi.org/10.1016/j.neucom.2018.06.065 (2018)

A semi-analytical model has been derived. The 1st order differential equation associated with the recovery of the left-hand side of the cell is solved to produce a time constant and time dependent voltage $V_{IN}(t)$. This voltage dependence is then applied to the right-hand side to give V_{PSP} (t). Note that the current-mirror transistors operate subthreshold for this model. An extension for the case of above threshold operation has also been produced (unpublished).

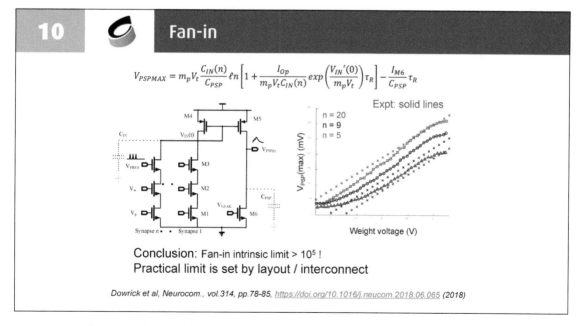

10 · Fan-in

$$V_{PSPMAX} = m_pV_t\frac{C_{IN}(n)}{C_{PSP}}\ell n\left[1 + \frac{I_{Op}}{m_pV_tC_{IN}(n)}\exp\left(\frac{V_{IN}'(0)}{m_pV_t}\right)\tau_R\right] - \frac{I_{M6}}{C_{PSP}}\tau_R$$

Expt: solid lines

n = 20
n = 9
n = 5

Conclusion: Fan-in intrinsic limit > 10^5 !
Practical limit is set by layout / interconnect

Dowrick et al, Neurocom., vol.314, pp.78-85, https://doi.org/10.1016/j.neucom.2018.06.065 (2018)

Comparison between the model and experimental results exploring fan-in. The transistors in the circuit operate below threshold.

11 Compact decision circuits (STDP)

Weight Increase, WI, Circuit Block, Output Buffers and SIFGNVM Device

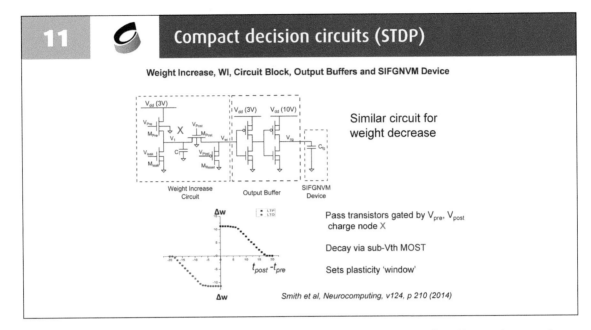

Weight Increase Circuit

Output Buffer

SIFGNVM Device

Similar circuit for weight decrease

Pass transistors gated by V_{pre}, V_{post} charge node X

Decay via sub-Vth MOST

Sets plasticity 'window'

Smith et al, Neurocomputing, v124, p 210 (2014)

Two compact circuits (one shown) detect the synaptic window and effect the STDP learning rule. The circuits operate dynamically and are compact in nature. It is shown how the block can be connected to an output buffer with sufficient drive to charge a floating gate device to increase the stored charge and hence the weight. A complementary block (not shown) can implement reduction of the weight.

12 How it works

WI Block Operation Pre-Post Spiking Event

- When a presynaptic spike occurs (V_{Pre})
  - V_1 is pulled up to 3V- $V_{TMpre}(V_1)$, C_1 charges via M_{pre}
  - C_1 Slowly discharges via sub-threshold M_{leak}
  - V_{post} triggers the sample/hold as some time, t after V_{pre}

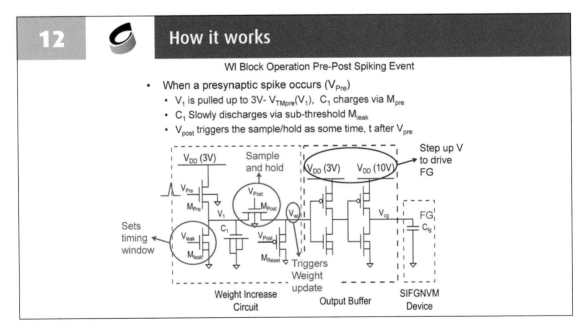

Weight Increase Circuit

Output Buffer

SIFGNVM Device

Details of cell operation. Subthreshold operation of the transistors allows time constants over a wide range of time.

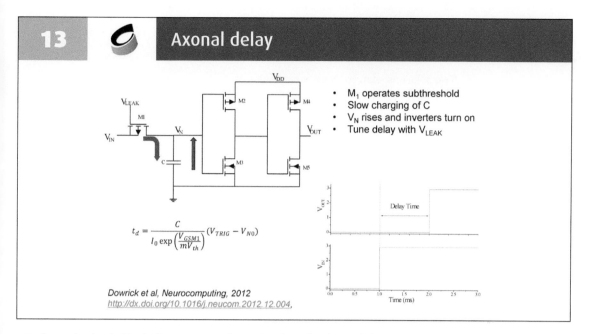

13 — Axonal delay

- M_1 operates subthreshold
- Slow charging of C
- V_N rises and inverters turn on
- Tune delay with V_{LEAK}

$$t_d = \frac{C}{I_0 \exp\left(\frac{V_{GSM1}}{mV_{th}}\right)}(V_{TRIG} - V_{N0})$$

Dowrick et al, Neurocomputing, 2012
http://dx.doi.org/10.1016/j.neucom.2012.12.004,

A dynamic circuit block that can introduce signal delay. The capacitor charges slowly via a MOSFET operating in subthreshold. The two inverters constitute a simple neuron. The voltage V_N rises slowly until the inverters switch, that is, the neuron fires. The delay time is tuneable over a wide range of time due to the exponential nature of the current through a MOSFET operating in subthreshold.

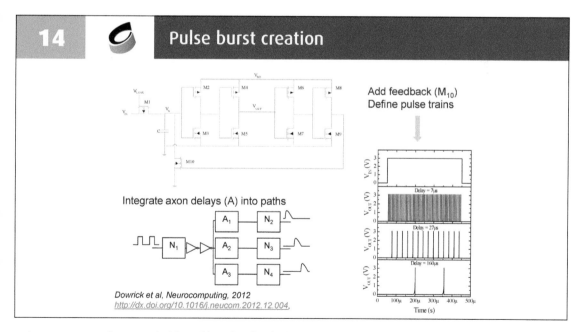

14 — Pulse burst creation

Add feedback (M_{10})
Define pulse trains

Integrate axon delays (A) into paths

Dowrick et al, Neurocomputing, 2012
http://dx.doi.org/10.1016/j.neucom.2012.12.004,

The concept can be extended by adding feedback allowing spike trains to be defined with a high degree of control. Different 'axon delays' can be built into the neuron blocks to allow closer adherence to biology.

15 Scaling

- Two solutions: sum voltages or sum currents

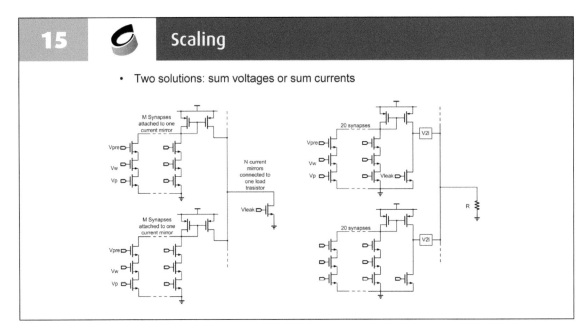

M ethods for scaling the circuit blocks are described.

16 Scaleability: easier to sum currents

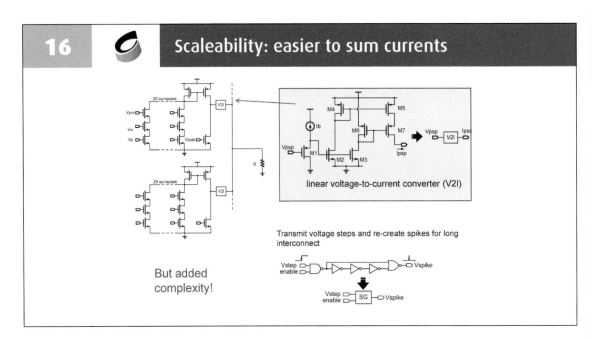

It is more robust to sum currents rather than voltages but this adds complexity. Transmitting spikes over long interconnect lines is to be avoided as they tend to be filtered to substrate. Voltage steps can be transmitted and spikes recovered at the next synapse. However, we are starting to add extra complexity and compromising packing density. Trade-offs to consider.

17 Scaling: circuit issues

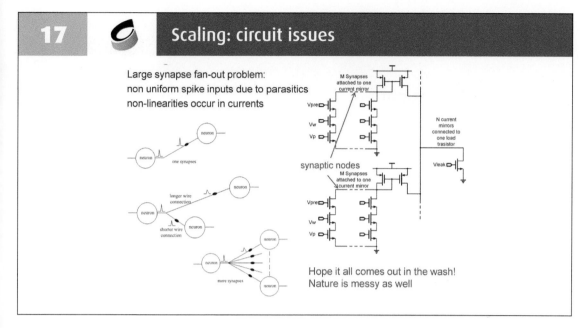

Large synapse fan-out problem:
non uniform spike inputs due to parasitics
non-linearities occur in currents

synaptic nodes

Hope it all comes out in the wash!
Nature is messy as well

A number of circuit related issues arise when circuits are scaled. VLSI implementations were beyond the scope of the project which was mainly concerned at the cell (building block) level. There is a hope that such issues are 'built-in' to the overall operation of the learning process.

18 Neurons with excitatory and inhibitory synapses

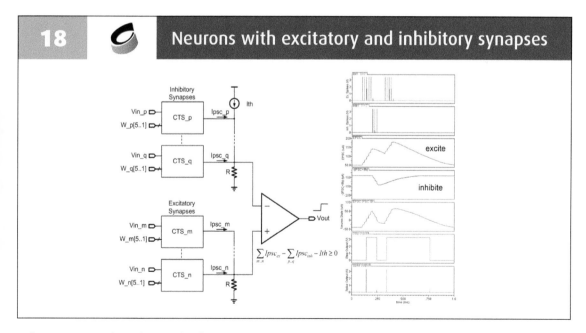

$$\sum_{m,n} Ipsc_{ex} - \sum_{p,q} Ipsc_{inh} - Ith \geq 0$$

There are a number of ways that both excitatory and inhibitory modes may be achieved – this shows one but it is expensive in term of packing density.

19 Programmable weights (I)

- ## Analog weight
  - Good: Continuous weight value, compact analog storage circuit
  - Bad: Inaccurate, require bias reference circuit and complex control circuit for high resolution, also require high voltage rail and undocumented
  - technology feature

- ## Digital weight
  - Good: accurate, mature digital memory technology, easy to program
  - Bad: discrete quantitative weight, require more space

20 Programmable weight (II)

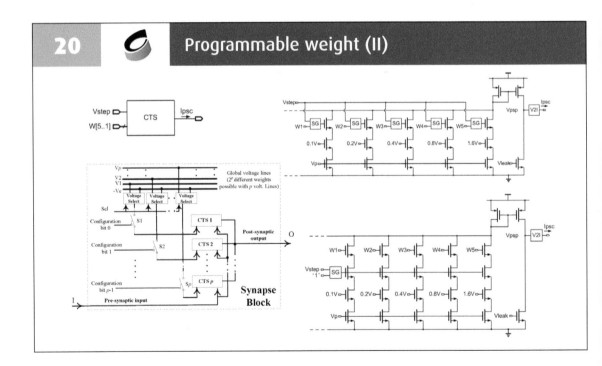

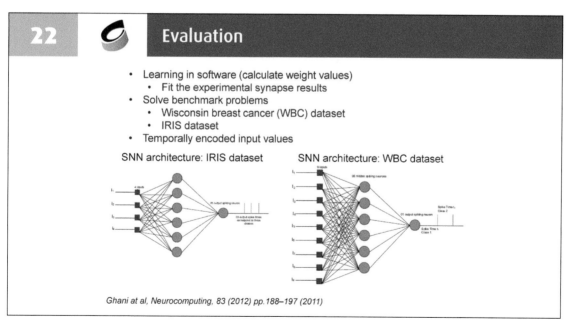

21 — Embrace: an alternative approach

- Network-on-chip address the issues of scalability and connectivity between components.
- Low-area/power spiking neuron cells with associated training provides neural computing capability.

- 2-dimensional array of interconnected neural tiles + I/O blocks.
- Neural tiles connected in North, East, South and West.
- Tile can be programmed to realise neuron-level functions.

Harkin et al, Int. Jnl of Reconfigurable Computing, doi:10.1155/2009/908740 (2009)

Slide courtesy of Jim Harkin

An alternative scaled architecture proposed by Jim Harkin of Ulster University. Not yet fabricated. There is a need for researchers with knowledge of building large systems to take over the work.

22 — Evaluation

- Learning in software (calculate weight values)
  - Fit the experimental synapse results
- Solve benchmark problems
  - Wisconsin breast cancer (WBC) dataset
  - IRIS dataset
- Temporally encoded input values

SNN architecture: IRIS dataset   SNN architecture: WBC dataset

Ghani at al, Neurocomputing, 83 (2012) pp.188–197 (2011)

Rather than build a large circuit to test the ideas (beyond our scope and expertise), synaptic dynamics derived from experimental results were modelled and built into higher-level algorithms.

The slide shows architectures for two well-known benchmark problems for NNs. The input and output are time-coded.

23   Circuits fabricated in AMS 0.35,mixed signal CMOS

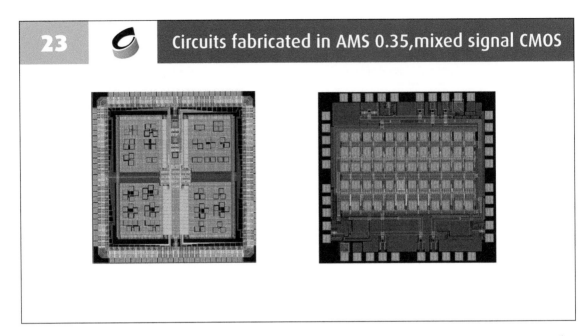

Two of the test chips fabricated in the course of the project. The one on the left contains a circuit to solve the XOR benchmark problem.

24   Astrocytes

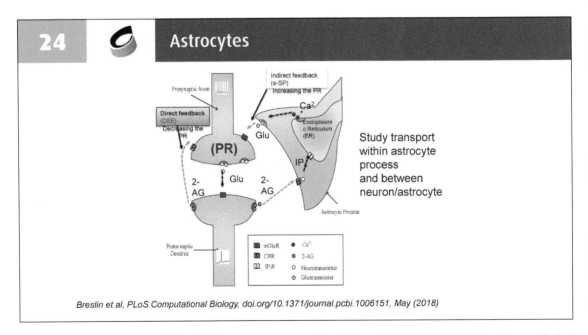

Study transport within astrocyte process and between neuron/astrocyte

Breslin et al, PLoS Computational Biology, doi.org/10.1371/journal.pcbi.1006151, May (2018)

The brain contains more glial cells than neurons! Astrocytes 'wrap around' synapses. Ionic pathways serve to 'monitor' neural activity and can 'detect' changing neural dynamics. The reference contains a detailed model of this ionic interaction and that shown in the next slide, demonstrates the recovery process.

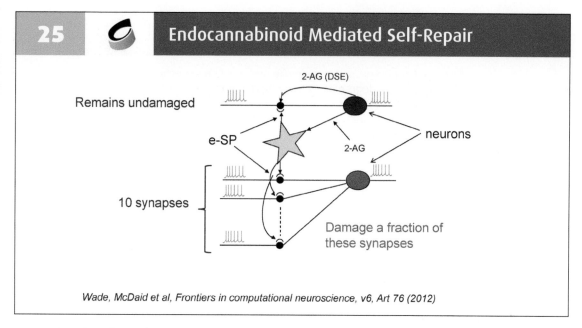

25 — Endocannabinoid Mediated Self-Repair

Wade, McDaid et al, Frontiers in computational neuroscience, v6, Art 76 (2012)

Example taken from the reference shown. The figure shows two neurons, the blue one having a single input (synapse) the red fed by 10 synapses.

The red indicates that damage has occurred to a fraction of the ten such that they do not transmit an input to the red neuron.

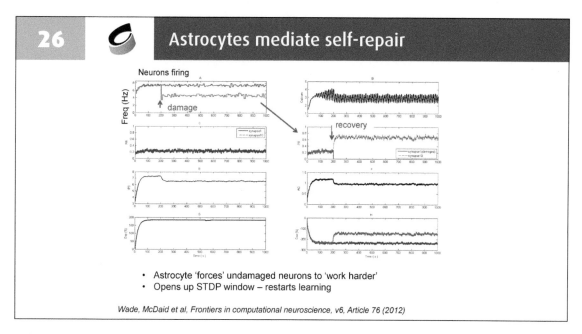

26 — Astrocytes mediate self-repair

- Astrocyte 'forces' undamaged neurons to 'work harder'
- Opens up STDP window – restarts learning

Wade, McDaid et al, Frontiers in computational neuroscience, v6, Article 76 (2012)

The damage occurs at 200 ms (top left-hand plot); the red neuron activity is decreased.

Right hand side, two down plot shows how the astrocyte has detected the fault and forced the undamaged neurons to 'work harder'. A further action is that the synaptic window is opened to restart the learning the process allowing the remaining synapses to be re-trained.

27 ## What we learnt..

- Can build compact analogue circuits that emulate aspects of biology with a degree of success (better than in software? – potentially much faster)
- Getting them to learn is another matter..
  - Need feedback
  - Weight update
  - Starts to get very complicated...
- A lot of redundancy once the circuit has 'learnt'
- Scaling soon results in a huge amount of interconnect

Need software/hardware combination – learning in software

28 ## Still some way to go before....

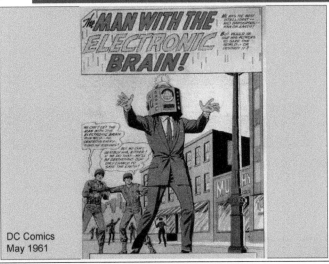

DC Comics
May 1961

29 ## Thanks to

J Harkin   *(jg.harkin@ulster.ac.uk)*

T Dowrick (PhD)
A Smith (PhD)
S Chen (PhD)
S Zhang (post-doc)
A Ghani (post-doc)

Funding: EPSRC, NAP, EPSRC-DTA awards, Dorothy Hodgkin scholarship

CHAPTER 05

Circuits for Bio-Potential Recording from the Brain

Srinjoy Mitra

University of Edinburgh

Advances in CMOS technology have made ultra-small electronic sensors become part of wearable and implantable devices. There are many challenges in designing components for such a device, however, the analog front-end circuits located right at the signal source is one of the most critical one to determine the overall performance. This tutorial will focus on low-power CMOS electronics for non-invasive and invasive bio-potential recoding from the central nervous system. Electro-encephalography (EEG) is the most popular method for non-invasive brain imaging but medical grade EEG devices are still bulky and requires frequent professional assistance. However, ambulatory EEG recording is of great interest both for healthcare and life-style applications. We will review the CMOS circuit design techniques to provide high quality EEG data with maximum patient comfort. Similarly, invasive neural recording has been a primary technique used by neurophysiologists to study brain function over many decades. It is only recently that high-density silicon neural probes have become available. The second part of the tutorial will focus on CMOS design methodologies related to such devices.

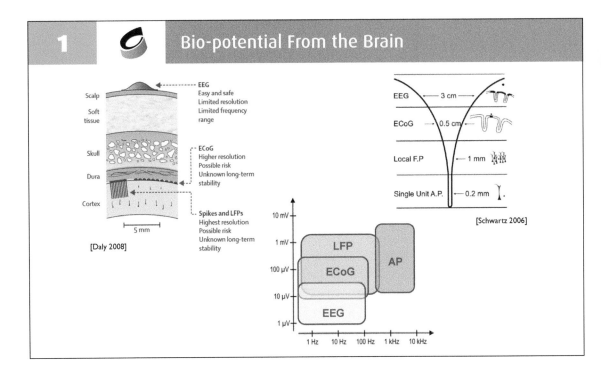

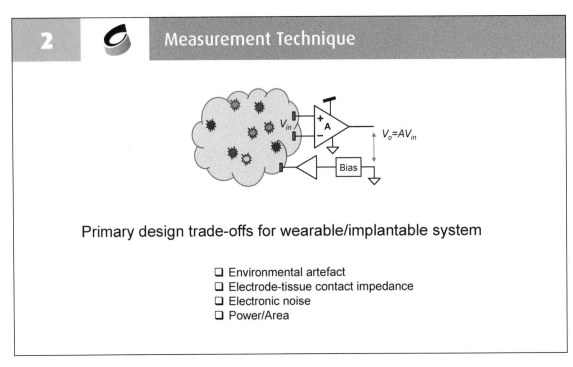

Basic bio-potential measurement technique and the associated circuits design trade-off.

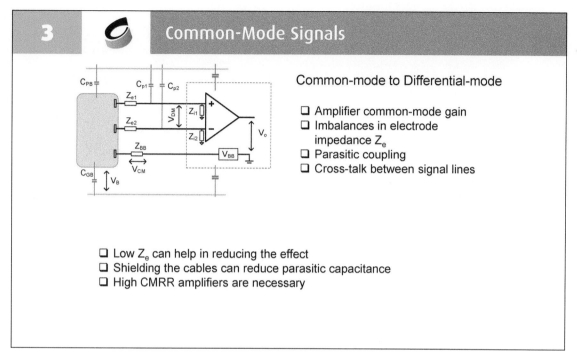

3 Common-Mode Signals

Common-mode to Differential-mode

❑ Amplifier common-mode gain
❑ Imbalances in electrode
   impedance Z_e
❑ Parasitic coupling
❑ Cross-talk between signal lines

❑ Low Z_e can help in reducing the effect
❑ Shielding the cables can reduce parasitic capacitance
❑ High CMRR amplifiers are necessary

Common-mode signal can arise from various environmental sources. They can convert to differential signal due to non-ideal components in the circuits.

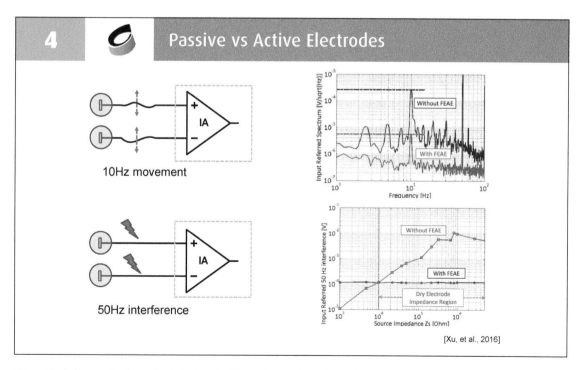

4 Passive vs Active Electrodes

10Hz movement

50Hz interference

[Xu, et al., 2016]

Practical demonstration of artefact reduction using active electrode

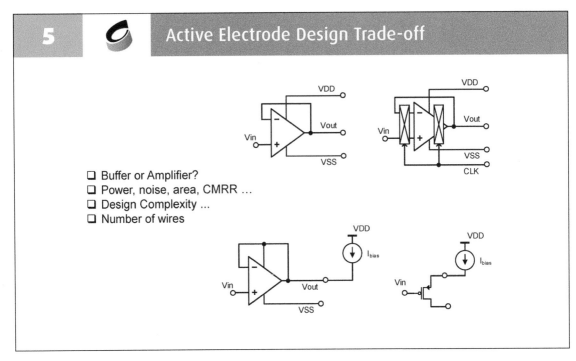

5 — **Active Electrode Design Trade-off**

- ❑ Buffer or Amplifier?
- ❑ Power, noise, area, CMRR …
- ❑ Design Complexity …
- ❑ Number of wires

Simple buffer, chopper stabilized buffer, 2-wire buffer, Single transistor buffer.

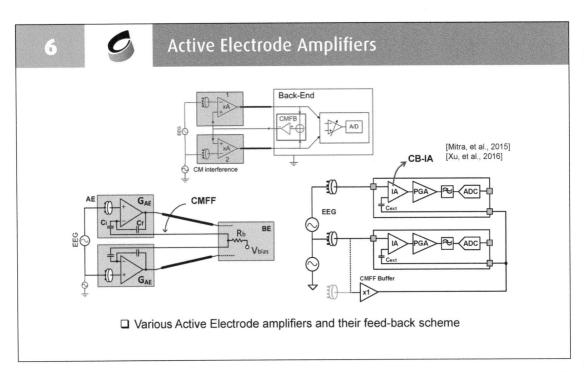

6 — **Active Electrode Amplifiers**

❑ Various Active Electrode amplifiers and their feed-back scheme

Active electrode amplifiers can help in power-noise trade-off but suffer from low CMRR. Back-end circuitry can be used to increase system CMRR. Common-mode feed-back, Common-mode feed-forward and digital active electrode.

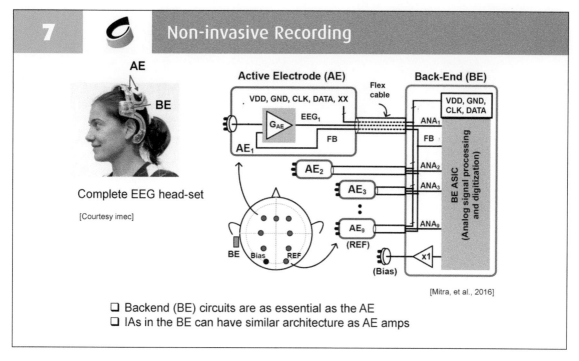

7 **Non-invasive Recording**

Complete EEG head-set

[Courtesy imec]

[Mitra, et al., 2016]

❑ Backend (BE) circuits are as essential as the AE
❑ IAs in the BE can have similar architecture as AE amps

Complete EEG head-set electronics used with dry electrode. Validated by multiple commercial vendors and used in clinical practice.

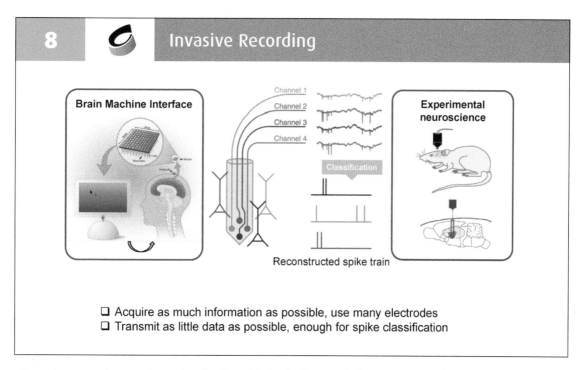

8 **Invasive Recording**

❑ Acquire as much information as possible, use many electrodes
❑ Transmit as little data as possible, enough for spike classification

Acquire more data: For better localization of individual neuron's firing pattern (spike classification).
Transmit less data: Transmission is power hungry, devices weigh more and also heat up the brain.

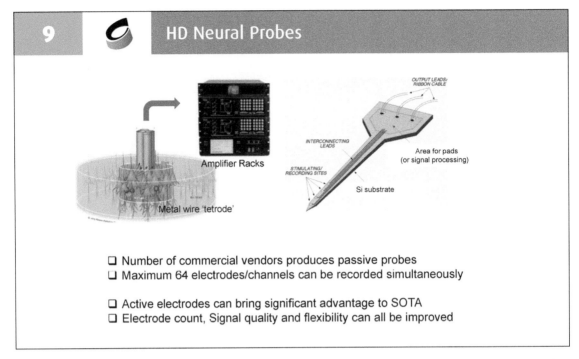

9   HD Neural Probes

❏ Number of commercial vendors produces passive probes
❏ Maximum 64 electrodes/channels can be recorded simultaneously

❏ Active electrodes can bring significant advantage to SOTA
❏ Electrode count, Signal quality and flexibility can all be improved

Neuroscientists have traditionally used metal electrodes (or tetrodes) and big rack amplifiers. Silicon probes, have created a whole new method electrophysiology recording. The need for even higher density neural probes have grown tremendously over the last decade.

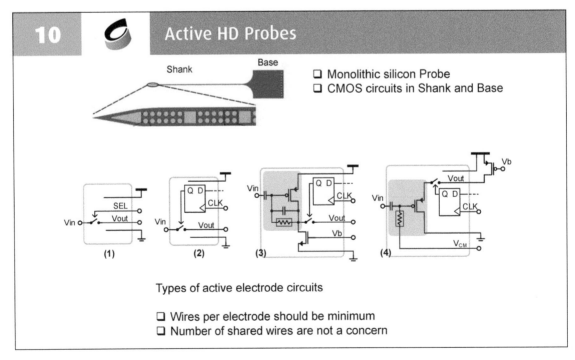

10   Active HD Probes

❏ Monolithic silicon Probe
❏ CMOS circuits in Shank and Base

Types of active electrode circuits

❏ Wires per electrode should be minimum
❏ Number of shared wires are not a concern

CMOS active probes are a new kind of electrophysiology tool, but needs to have unusually form factor compared to a traditional silicon IC. 1,2 has small electrode size. But does not use 'active electrode' features. 3 has gain, 4 is a buffer.

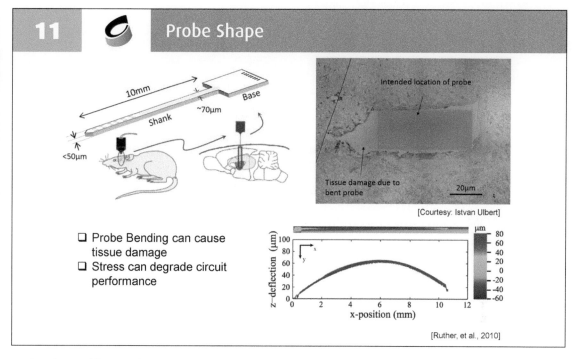

11   **Probe Shape**

- ❏ Probe Bending can cause tissue damage
- ❏ Stress can degrade circuit performance

[Courtesy: Istvan Ulbert]

[Ruther, et al., 2010]

Ultra thin and long probe can cause significant bending and stress (on silicon).

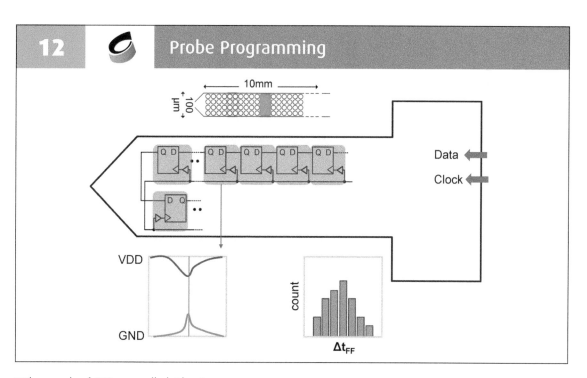

12   **Probe Programming**

Thousands of DFF manually laid out.
Serial connection, inefficient clock/data tree.
Simultaneous transition can create huge current surge.
DFF delay can have significant spread resulting in SU/HOLD violation.

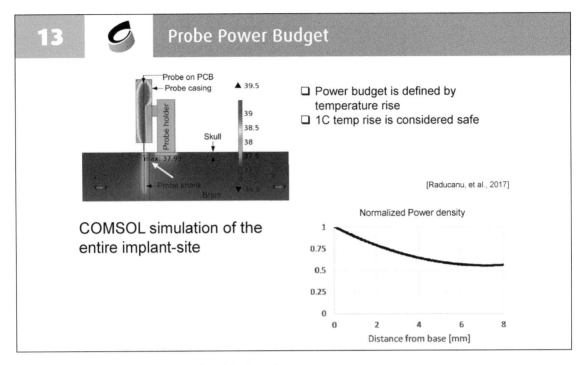

13 — Probe Power Budget

❏ Power budget is defined by temperature rise
❏ 1C temp rise is considered safe

[Raducanu, et al., 2017]

COMSOL simulation of the entire implant-site

Entire head-stage should be considered for heat flow.
Power dissipated for data communication can flow down to the shank.

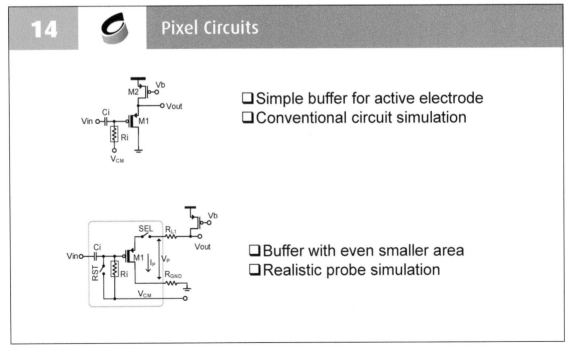

14 — Pixel Circuits

❏ Simple buffer for active electrode
❏ Conventional circuit simulation

❏ Buffer with even smaller area
❏ Realistic probe simulation

Voltage drops over RL1 and RGND can be significant.
The drop varies depending on the position of the pixel.
Individual pixel bias and CMRR will be affected.
Power dissipation within a pixel = VPIP.

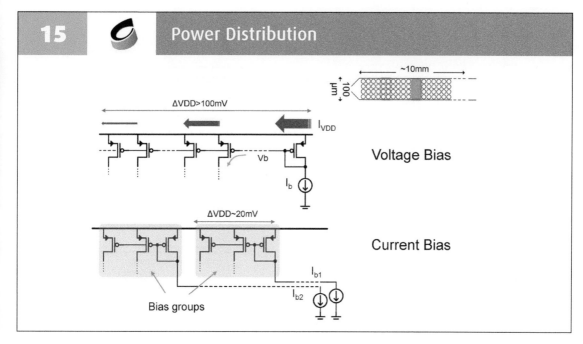

15 ⬤ **Power Distribution**

ΔVDD>100mV

I_{VDD}

Voltage Bias

Vb

I_b

ΔVDD~20mV

Current Bias

Bias groups

I_{b1}
I_{b2}

VDD keeps changing over the length of the probe
   Resistive drop over VDD creates difference from Vb.
Vb itself changes due to massive leakage
   Current bias and Vb regeneration solves some

problem.
   Wiring complexity limits number of groups
   Still significant ΔVDD.

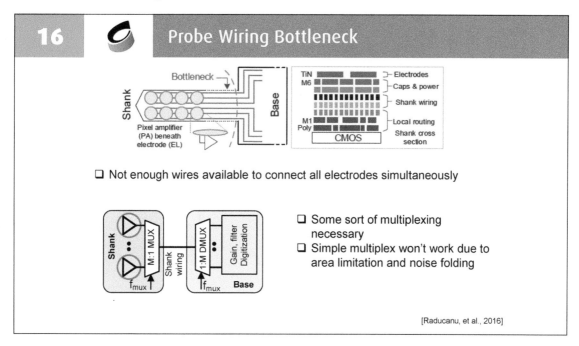

16 ⬤ **Probe Wiring Bottleneck**

Bottleneck

Shank

Base

Pixel amplifier
(PA) beneath
electrode (EL)

TiN
M6 ── Electrodes
── Caps & power
── Shank wiring
M1
Poly
── Local routing
   Shank cross
   section
CMOS

❑ Not enough wires available to connect all electrodes simultaneously

Shank

M:1 MUX

Shank
wiring

1:M DMUX

Gain, filter
Digitization

f_{mux} f_{mux} **Base**

❑ Some sort of multiplexing
   necessary
❑ Simple multiplex won't work due to
   area limitation and noise folding

[Raducanu, et al., 2016]

Even for high density probes, not all electrodes can be recorded simultaneously. This is due to the number of signal wires that can fit within the narrow shank cross-section. This bottleneck can be removed

with some multiplexing. Extremely low area per pixel doesn't allow for anti-aliasing filter, leading to noise-folding.

17 — Probe Examples

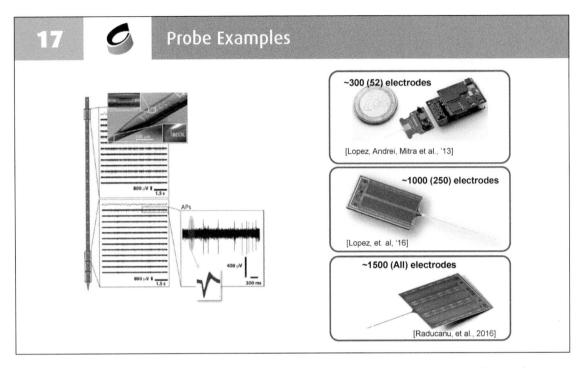

Extensive post processing done on each probe. TiN electrode deposition, Thinning, Deep silicon etching, etc. Various generations of probes designed.

18 — Probe Performance

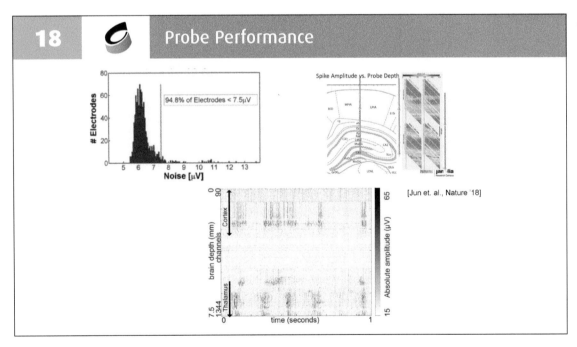

Extremely unusual ASIC with >500k analog transistors. Circuit noise performance show a high degree of conformity..

Probes used by multiple labs around the world.

High profile publications.

Rastar plot with >1000 electrodes, never seen before.

19 Novel Neuroscience

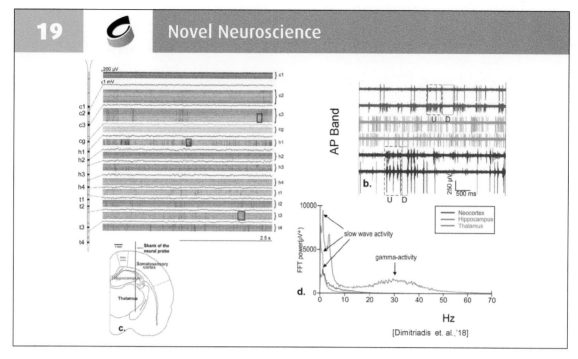

[Dimitriadis et. al.,'18]

An entire cortical column can now be recorded giving new insight about the brain. Correlation between activity pattern in different layers demonstrated.

20 Data Compression

- ❑ Action Potential bandwidth: 500Hz-10kHz
- ❑ 1000 channels: data rate ~500Mbps
- ❑ Data compression is absolutely necessary for large systems
- ❑ Compressed sensing (CS) takes advantage of sparse data

[Zhang, Mitra et al., '14; '16]

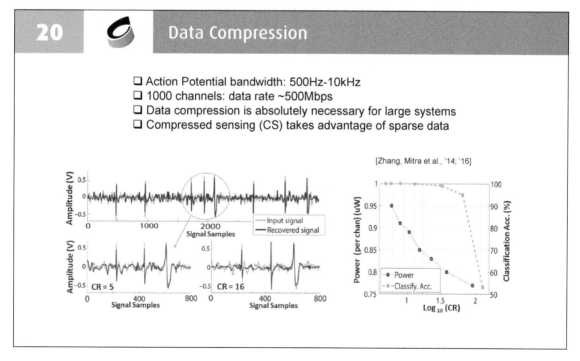

With such high density probe and simultaneous recording of all channels, data compression is absolutely necessary. CS is a very suitable compression technique and can be adapted for chronic recording, without any loss in classification accuracy.

21 References

1. B. C. Raducanu, R. F. Yazicioglu, C. M. Lopez, M. Ballini, J. Putzeys, S. Wang, A. Andrei, V. Rochus, M. Welkenhuysen, N. van Helleputte, S. Musa, R. Puers, F. Kloosterman, C. van Hoof, R. Fiáth, I. Ulbert, and S. Mitra, "Time Multiplexed Active Neural Probe with 1356 Parallel Recording Sites," *Sensors*, vol. 17, no. 10, p. 2388, 2017.

2. A. S. Herbawi, F. Larramendy, T. Galchev, T. Holzhammer, B. Mildenberger, O. Paul, and P. Ruther, "CMOS-based neural probe with enhanced electronic depth control," *2015 Transducers - 2015 18th Int. Conf. Solid-State Sensors, Actuators Microsystems, TRANSDUCERS 2015*, pp. 1723–1726, 2015.

3. T. Torfs, A. A. A. Aarts, M. A. Erismis, J. Aslam, R. F. Yazicioglu, K. Seidl, S. Herwik, I. Ulbert, B. Dombovari, R. Fiáth, B. P. Kerekes, R. Puers, O. Paul, P. Ruther, C. Van Hoof, and H. P. Neves, "Two-dimensional multi-channel neural probes with electronic depth control," *IEEE Trans. Biomed. Circuits Syst.*, vol. 5, no. 5, pp. 403–412, 2011.

4. K. Seidl, M. Schwaerzle, I. Ulbert, H. P. Neves, O. Paul, and P. Ruther, "CMOS-based high-density silicon microprobe arrays for electronic depth control in intracortical neural recording-characterization and application," *J. Microelectromechanical Syst.*, vol. 21, no. 6, pp. 1426–1435, 2012.

5. G. Buzsáki, E. Stark, A. Berényi, D. Khodagholy, D. R. Kipke, E. Yoon, and K. D. Wise, "Tools for probing local circuits: High-density silicon probes combined with optogenetics," *Neuron*, vol. 86, no. 1, pp. 92–105, 2015.

6. J. Scholvin, C. G. Fonstad, and E. S. Boyden, "Scaling models for microfabricated in vivo neural recording technologies," *Int. IEEE/EMBS Conf. Neural Eng. NER*, pp. 181–185, 2017.

7. W. Bruce and J. G. Webster, "Communications On the Reduction of Interference Due to Common Mode Voltage in Two-Electrode Biopotential Amplifiers," *IEEE Trans. Biomed. Eng.*, vol. BME-33, no. 11, pp. 1043–1046, 1986.

8. E. M. Spinelli, M. A. Mayosky, and R. Pallás-Areny, "A practical approach to electrode-skin impedance unbalance measurement," *IEEE Trans. Biomed. Eng.*, vol. 53, no. 7, pp. 1451–1453, 2006.

9. R. Pallás-Areny, C. Josep, and R. Javier, "An Improved Buffer for Bioelectric Signals," *IEEE Trans. Biomed. Eng.*, vol. 36, no. 4, pp. 490–493, 1989.

10. A. C. Metting van Rijn, A. Peper, and C. A. Grimbergen, "Isolation_mode_rejection. pdf," *IEEE Trans. Biomed. Eng.*, vol. 38, no. 11, pp. 1154–1156, 1991.

11. A. C. Metting van Rijn, A. Peper, and C. A. Grimbergen, "High-quality recording of bioelectric events - Part 1 Interference reduction, theory and practice," *Medical & Biological Engineering & Computing*, vol. 28, no. 5. pp. 389–397, 1990.

12. T. D. Towers, "High Input-impedance Amplifier Circuits," vol. 1968, no. July, pp. 197–201, 1968.

13. T. Degen and H. Jackel, "A pseudodifferential amplifier for bioelectric events with DC-offset compensation using two-wired amplifying electrodes," *IEEE Trans. Biomed. Eng.*, vol. 53, no. 2, pp. 300–310, 2006.

14. D. E. Wood, D. J. Ewins, and W. Balachandran, "Comparative analysis of power-line interference between two- or three-electrode biopotential amplifiers," *Med. Biol. Eng. Comput.*, vol. 33, no. 1, pp. 63–68, 1995.

15. J. Xu, R. F. Yazicioglu, B. Grundlehner, P. Harpe, K. A. A. Makinwa, and C. Van Hoof, "A 160 uw 8-channel active electrode system for EEG monitoring," *IEEE Trans. Biomed. Circuits Syst.*, vol. 5, no. 6, pp. 555–567, 2011.

16. J. C. Huhta and J. G. Webster, "60-Hz Interference in Electrocardiography," *IEEE Trans. Biomed. Eng.*, vol. BME-20, no. 2, pp. 91–101, 1973.

17. M. F. Chimeno and R. Pallàs-Areny, "A comprehensive model for power line interference in biopotential measurements," *IEEE Trans. Instrum. Meas.*, vol. 49, no. 3, pp. 535–540, 2000.

18. A. B. Usakli, "Improvement of EEG signal acquisition: An electrical aspect for state of the Art of front end," *Comput. Intell. Neurosci.*, vol. 2010, 2010.

19. R. Pallás-Areny, "Interference-rejection characteristics of biopotential amplifiers: A comparative analysis," *IEEE Trans. Biomed. Eng.*, vol. 35, no. 11, pp. 953–959, 1988.

20. C. M. Lopez, J. Putzeys, B. C. Raducanu, G. Student, M. Ballini, S. Wang, A. Andrei, V. Rochus, R. Vandebriel, S. Severi, C. Van Hoof, S. Musa, and N. Van, "A Neural Probe with up to 966 Electrodes and up to 384 Configurable Channels in 0.13 μm SOI CMOS," 2016.

21. J. Zhang, S. Mitra, Y. Suo, A. Cheng, T. Xiong, F. Michon, M. Welkenhuysen, F. Kloosterman, P. S. Chin, S. Hsiao, T. D. Tran, F. Yazicioglu, and R. Etienne-Cummings, "A closed-loop compressive-sensing-based neural recording system," *J. Neural Eng.*, vol. 12, no. 3, pp. 1–17, 2015.

22. M. Guermandi, R. Cardu, E. F. Scarselli, and R. Guerrieri, "Active electrode IC for EEG and electrical impedance tomography with continuous monitoring of contact impedance," *IEEE Trans. Biomed. Circuits Syst.*, vol. 9, no. 1, pp. 21–33, 2015.

23. C. M. Lopez, A. Andrei, S. Mitra, M. Welkenhuysen, W. Eberle, C. Bartic, R. Puers, R. F. Yazicioglu, and G. G. E. Gielen, "An implantable 455-active-electrode 52-channel CMOS neural probe," *IEEE J. Solid-State Circuits*, vol. 49, no. 1, pp. 248–261, 2014.

24. J. Xu, S. Mitra, A. Matsumoto, S. Patki, C. Van Hoof, K. A. A. Makinwa, and R. F. Yazicioglu, "A wearable 8-channel active-electrode EEG/ETI acquisition system for body area networks," *IEEE J. Solid-State Circuits*, vol. 49, no. 9, pp. 2005–2016, 2014.

22 **Acknowledgement**

- ❑ Medical Electronics Group (imec, Leuven, Belgium)
- ❑ Ralph Etienne Cummings (Johns Hopkins University, Baltimore, USA)
- ❑ IMNS, University of Edinburgh

Advances in Scalable Implantable Stimulation Systems for Neuroprostheses using Networked ASICs

Xiao Liu

University College London, UK

D amage or degradation to the central and peripheral nervous systems due to injury or disease results in loss of neural function in various parts of the body. Neuroprostheses may assist in partial restoration of function and mobility using neurostimulation. This tutorial describes recent advances in scalable implantable stimulation systems using networked ASICs. It discusses how they can meet the ever-growing demand for high-density neural interfacing and long-term reliability. A detailed design example of an implantable (inductively linked) scalable stimulation system for restoring lower limb functions in paraplegics after spinal cord injury is presented.

1 Outline

1. Introduction on neural stimulation and applications

2. Different system architectures for implantable multi-channel neurostimulation

3. Challenges for realizing a scalable implantable neurostimulation system (a network of small implants!)

4. A design example using networked application-specific integrated circuits (ASICs)

5. Conclusion

2 The Nervous System

Damage or degradation to the central and peripheral nervous systems due to the injury or disease may result in loss of neural functions in various parts of the body.

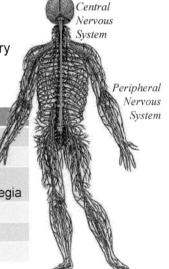

Central Nervous System

Peripheral Nervous System

CNS	PNS
Cochlear	Hand grasp
Retinal	Foot drop
Parkinson's disease	Spinal cord injury: Incontinence, Paraplegia
Epilepsy	Epilepsy
Obstructive Sleep Apnea	Depression
Pain	Pain

3   Implantable Technologies for Neuroprosthesis

The use of implantable technologies to restore lost functions of neurologically impaired patients has become a well-established treatment modality.

When an implanted device is used for neuroprosthesis, it
delivers electrical stimulus to the neuromuscular system →Neural stimulation
and/or
records neural signals from the neuromuscular system. →Neural recording

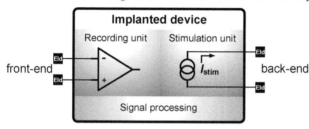

Interact with the neural environment via an intimate **electronics-tissue interface**.

4   Market

The global market for implantable neuroprosthetic devices is expected to reach USD 10.48 billion by 2022 from USD 5.84 billion in 2017 at a compound annual growth rate of 12.4%.

The market is dominated by five applications:
SCS: spinal cord stimulation
CI: cochlear implants
DBS: deep brain stimulation
SRS: sacral root stimulation
VNS: vagal nerve stimulation

["Neuroporsthetics market by type, techniques, applications - global forcast to 2022," market report by MarketsandMarkets, 2018
Pikov, "Global market for implanted neuroprostheses," book chapter in *Implantable Neuroprostheses for Restoring Function*, 2015]

5 The Control Chart

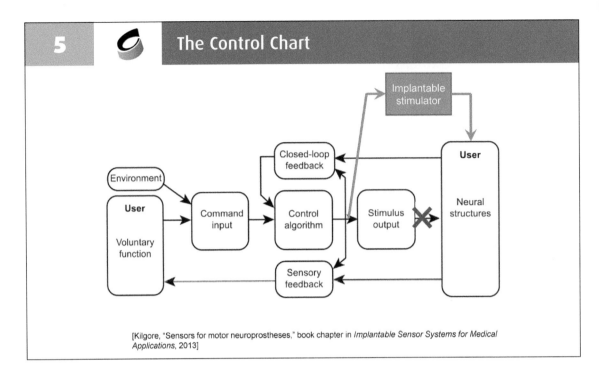

[Kilgore, "Sensors for motor neuroprostheses," book chapter in *Implantable Sensor Systems for Medical Applications*, 2013]

6 Recent Advance

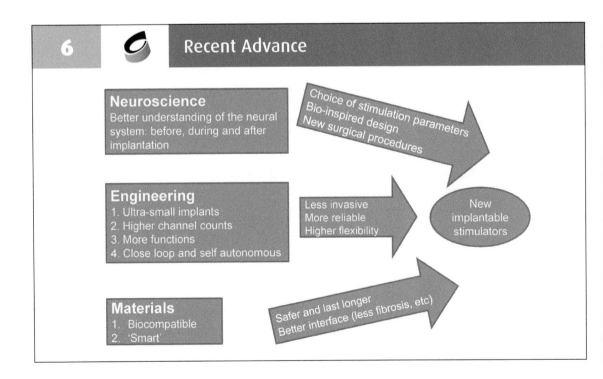

7 Small Implants and Many Channels

Spinal cord stimulation for controlling lower limbs
(inc. leg movement and bladder control)

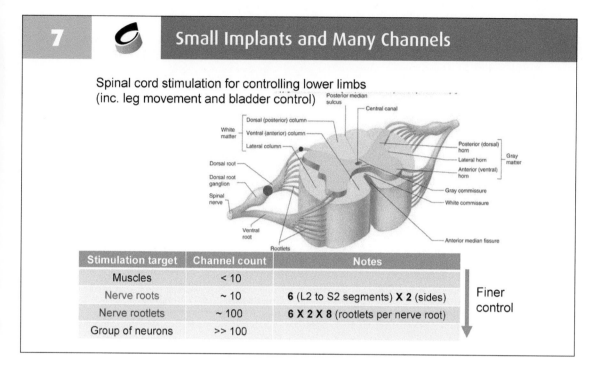

Stimulation target	Channel count	Notes	
Muscles	< 10		
Nerve roots	~ 10	**6** (L2 to S2 segments) **X 2** (sides)	Finer control
Nerve rootlets	~ 100	**6 X 2 X 8** (rootlets per nerve root)	
Group of neurons	>> 100		

8 Multi-Channel Implant Systems

What can be done in order to increase the number of channels?

1. One implant with lots of channels. There will be a limit.

2. Multiple implants. Each individual one can work alone. (Examples: USC's BION System)

3. Multiple small devices which are managed by a central implant, known as the hub (Example: UC Berkley's Neural Dust, CWRU's Networked Neuroprosthetic System and UCL's Active Book System)

Implant network

9 USC's BION System

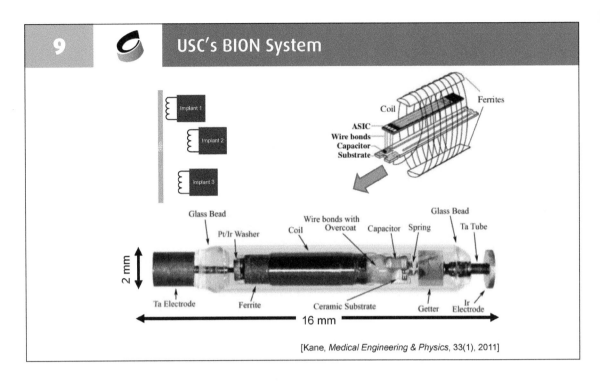

[Kane, *Medical Engineering & Physics*, 33(1), 2011]

10 CWRU's Networked Neuroprosthetic System

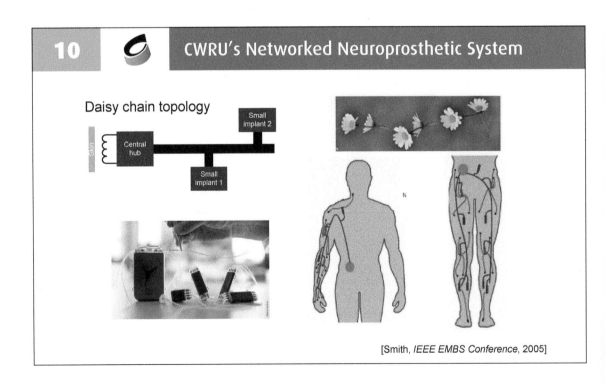

[Smith, *IEEE EMBS Conference*, 2005]

11   UC Berkley's Neural Dust

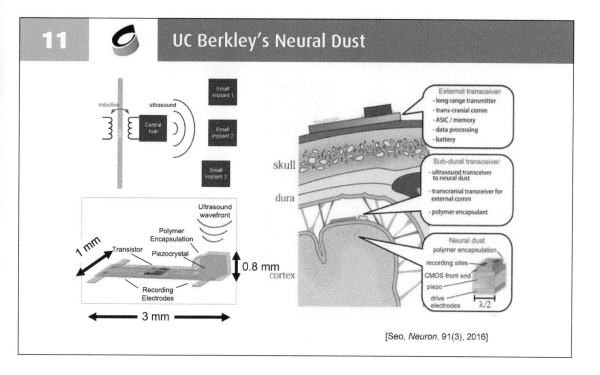

[Seo, *Neuron*, 91(3), 2016]

DESIGN CHALLENGE: LEAD COUNT

12   Implantable Electrodes

Cochlear implant electrodes

Michigan Electrodes
[Wise, *Proceedings of IEEE*, 96(7), 2008]

Utah Electrode Array
[Kim, *Biomedical Microdevices*, 11(2), 2009]

Carbon Nanotube Microelectrode Array
[Wang, *Nano Letters*, 6(9), 2006]

13 — Electrode Driver

Implanted device

$2n$ interconnecting wires needed!

De-multiplexer for steering stimulus current to the desirable electrode

An ASIC is embedded in the electrode mount. It does i) stimulus regeneration and ii) demultiplexing.

[Liu, IEEE TBioCAS, 6(3), 2012]

14 — Addressing Individual Electrodes

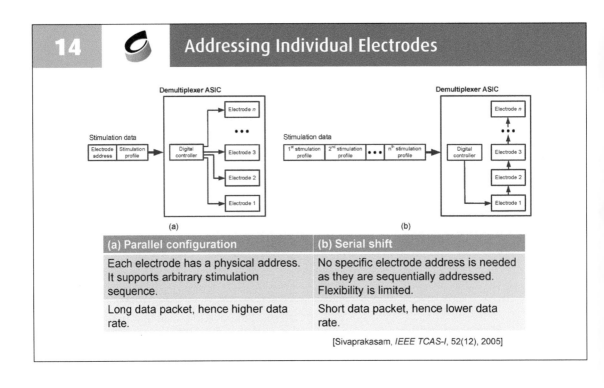

(a)   (b)

(a) Parallel configuration	(b) Serial shift
Each electrode has a physical address. It supports arbitrary stimulation sequence.	No specific electrode address is needed as they are sequentially addressed. Flexibility is limited.
Long data packet, hence higher data rate.	Short data packet, hence lower data rate.

[Sivaprakasam, *IEEE TCAS-I*, 52(12), 2005]

DESIGN CHALLENGE: LARGE BLOCKING CAPACITORS

15 Blocking Capacitors (Discrete)

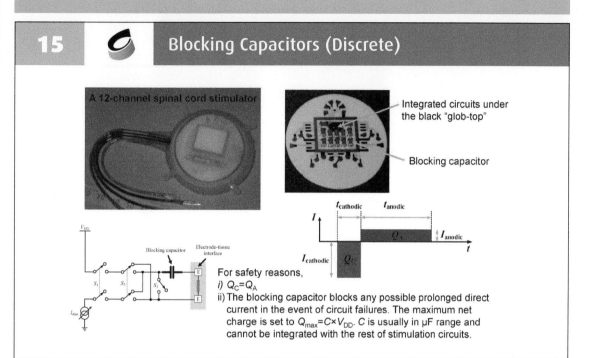

Integrated circuits under the black "glob-top"

Blocking capacitor

For safety reasons,

i) $Q_C = Q_A$

ii) The blocking capacitor blocks any possible prolonged direct current in the event of circuit failures. The maximum net charge is set to $Q_{max} = C \times V_{DD}$. C is usually in µF range and cannot be integrated with the rest of stimulation circuits.

16 High-Frequency Current-Switching (HFCS)

Continuous stimulation pulse (tens of µs to several ms)

$$C = I_{stim} \frac{\Delta t}{\Delta V}$$

$$C \propto \Delta t$$

I_{stim}

I_1

I_2

Two complementary high-frequency pulses

$$1\mu F = 1mA \times \frac{1ms}{1V}$$

$$50pF = 1mA \times \frac{50ns}{1V}$$

$$50pF = 1mA \times \frac{50ns}{1V}$$

[Liu, *IEEE TBioCAS*, 2(3), 2008]

17 — Fail-Safe Stimulator Output Stage (with Integrated Blocking Capacitor)

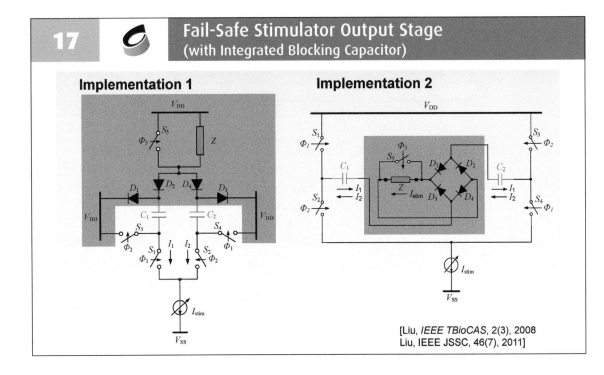

[Liu, *IEEE TBioCAS*, 2(3), 2008
Liu, IEEE JSSC, 46(7), 2011]

DESIGN CHALLENGE: HERMETICITY

18 — Hermeticity of Implant Package (I)

Hermeticity :
'The state or condition of being airtight'
-Webster's New Collegiate Dictionary

'Sealed so that the object is airtight'
-Microelectronics Packaging Handbook

Three out of four device failures were attributed to moisture ingression into the titanium receiver-stimulator packaging of an implant through the feedthrough.
- An auditory reliability report in 2006.

19 Hermeticity of Implant Package (II)

All materials and all welded or joined assemblies leak to some degree, whether by permeation through the bulk material or along a discontinuity path.

The degree and measure of hermeticity are a function of:
- material choice
- final seal design
- fabrication process and practice
- Use environment

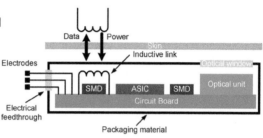

20 Hermeticity of Implant Package (III)

A commonly cited limit for acceptable humidity inside a hermetic implant package is 5000ppm which corresponds to a relative humidity of 8.1%. If too much moisture has penetrated the implant package, it may cause corrosion and failure of electronic circuits.

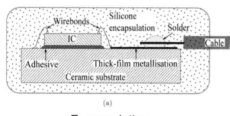

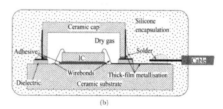

Encapsulation   Micro-package / seal

[Schuettler, *IEEE EMBC conference*, 2011
Vanhoestenberghe, *Journal of Neural Engineering*, 10(3), 2013]

21 Measurement Methods

Gross leak tests (for leak rates > 1×10^{-5} atm-cc/sec):
Bubble test, vapor detection test, weight gain test, measure the deflection after pressure change, dye penetrant test, residual gas analysis

Fine leak tests (for leak rates < 1×10^{-5} atm-cc/sec):
- Helium-leak detection
- Optical detection

Most fine leak test methods use helium tracer gas
- Inert
- Small molecule
- Easy to detect
- Low levels in background atmosphere

22 Helium Leak Test

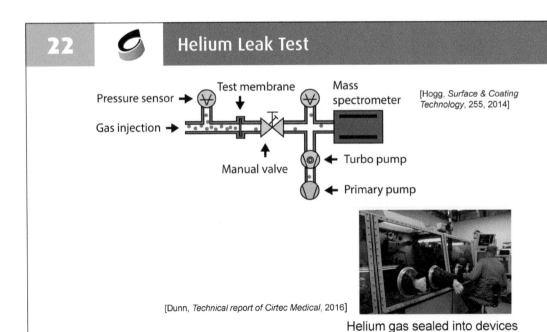

[Hogg, *Surface & Coating Technology*, 255, 2014]

[Dunn, *Technical report of Cirtec Medical*, 2016]

Helium gas sealed into devices

23 Limitation of Gas Leak Test

The gas leak test only works for implants with an inner cavity larger than 50 mm^3. However, ultra-small implanted devices with inner cavities of a few mm^3 or even sub-mm^3 mean that the conventional packaging methods will either not work or become too hard to validate.

Humidity sensors may be used instead to directly assess the in-package humidity, especially integrated humidity sensors as they can be integrated with the rest of implant circuits and bear little overhead to the entire implant size.

[Vanhoestenberghe, *Artificial Organs*, 35(3), 2011]

24 Different Integrated Humidity Sensors

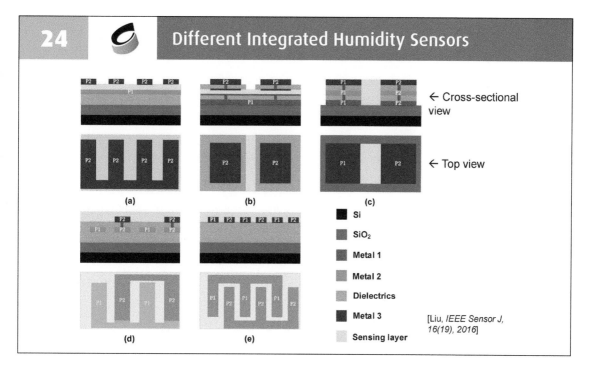

[Liu, *IEEE Sensor J, 16(19), 2016*]

25   Typical Types of Readout Circuits for Capacitive RH Sensors

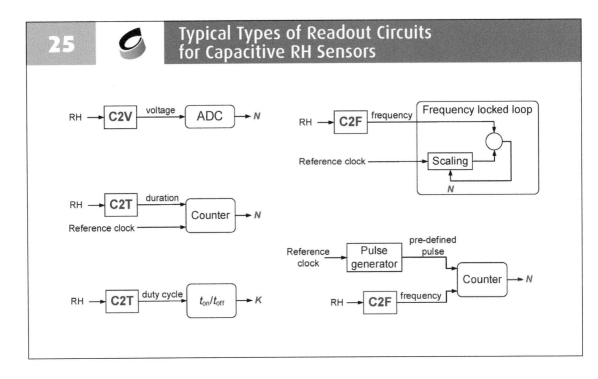

DESIGN CHALLENGE: LOW-POWER OPERATION

26   Power Consumption

In an implantable stimulator, power is supplied by an implanted battery or wireless link.

The power consumption of an active electrode is

$$P_{AE} = V_{supply}(I_{stim} + I_{other})$$

V_{supply}: the supply voltage to the stimulator ASIC,

I_{stim}: the stimulus current amplitude, mainly dictated by the physiological response of the neural tissue

I_{other}: any other current taken from V_{supply}, inc. the current necessary for bias circuits, monitoring circuits and signal processing circuits.

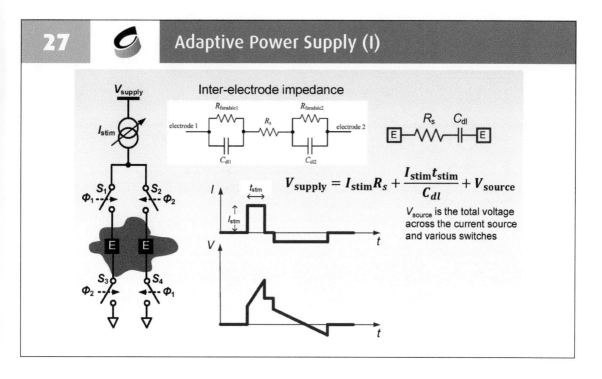

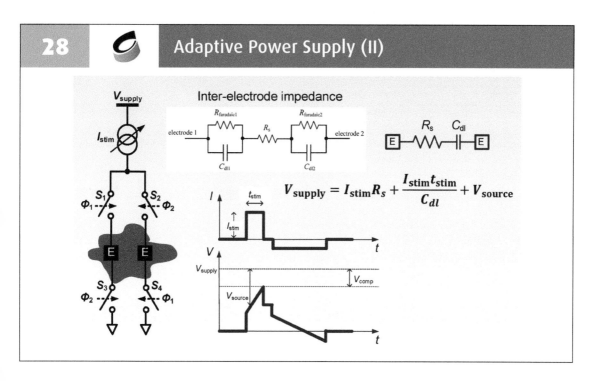

29 Adaptive Power Supply (III)

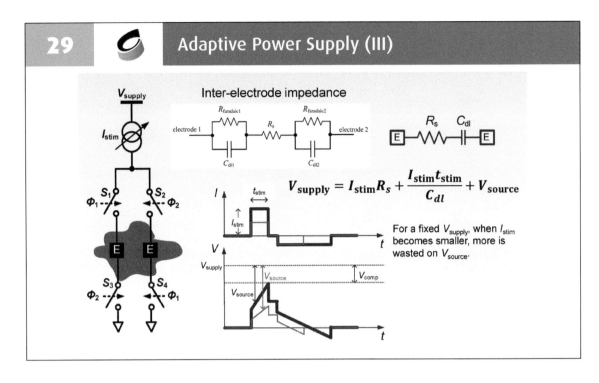

$$V_{\text{supply}} = I_{\text{stim}}R_s + \frac{I_{\text{stim}}t_{\text{stim}}}{C_{dl}} + V_{\text{source}}$$

For a fixed V_{supply}, when I_{stim} becomes smaller, more is wasted on V_{source}.

30 Adaptive Power Supply (IV)

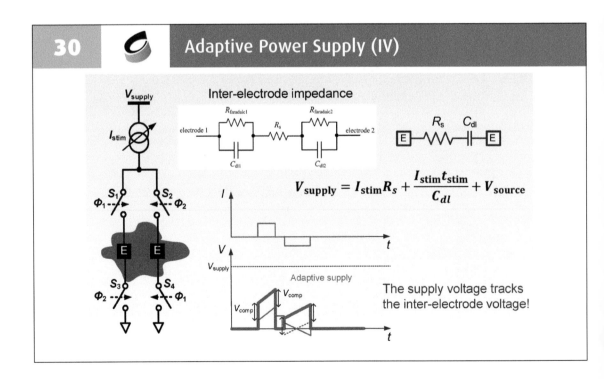

$$V_{\text{supply}} = I_{\text{stim}}R_s + \frac{I_{\text{stim}}t_{\text{stim}}}{C_{dl}} + V_{\text{source}}$$

The supply voltage tracks the inter-electrode voltage!

DESIGN CHALLENGE: HEAT DISSIPATION

31 Over-Temperature in Tissue

Conventional implanted devices, such as pacemakers, are relatively large. The electronic circuits are usually enclosed in a metal casing. The power dissipation density from the surface of the ASIC becomes higher as implanted devices becomes smaller.

Excessive heat exceeding the threshold for thermal damage to the nervous system can cause i) vascular effects, ii) direct damage, and iii) functional effects.

Animal studies have shown that a maximum of 40 mW/cm² chronic heat flux can be tolerated for implanted devices, as this has been shown to lead to less than 1 °C temperature increase of the surrounding tissue.

Cumulative equivalent minutes at 43°C (CEM43) is the accepted metric for thermal dose assessment.

[Davies, *ASAIO J.*, 40(3), 1994
Yarmolenko, *Int J Hyperthermia*, 27(4), 2011]

DESIGN EXAMPLE: ACTIVE BOOK SYSTEM

32 Passive Book

Book Electrodes have been implanted in more than 2500 patients.

10 mm

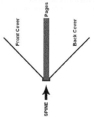

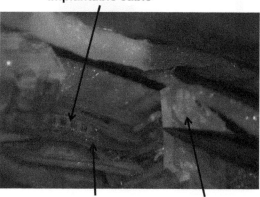

Implantable cable

Nerve rootlets   **Book Electrodes**

33 Active Book

By embedding a small stimulator ASIC into the spine region, a passive BOOK is turned into an active BOOK.

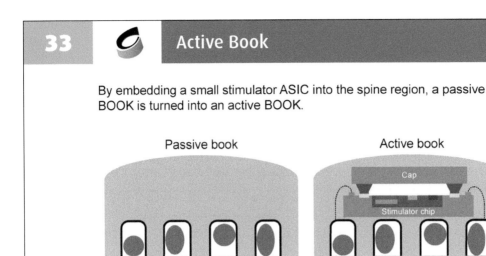

Passive book   Active book

- Nerve root
- *Book* structure (in silicone)
- Core silicon circuits
- Seal ring
- Temperature sensor

- Voltage sensor
- Humidity sensor
- U-shape electrode
- digital circuits
- Bonding pad
- Flexible tracks for bonding

34 Active Book Implant System

An implant system for restoring lower body function to patients with paraplegia after severe SCI by targeting stimulation of the lumbo-sacral nerve roots in the human cauda equina. It can enable exercising such as cycling and rowing, to control the bladder, and to improve bowel evacuation.

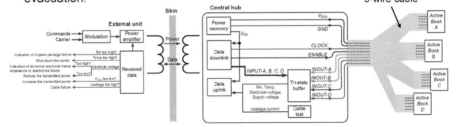

The total number of stimulation channels can be easily scaled up by having more slots in a Book and having more Books.
The total No. of channels = No. of Active Books × No. of slots per Book

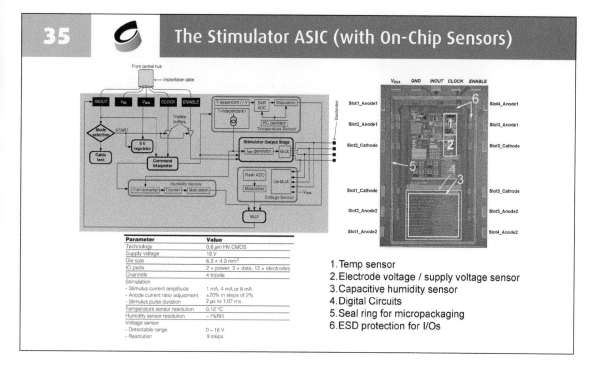

35   The Stimulator ASIC (with On-Chip Sensors)

Parameter	Value
Technology	0.6 µm HV CMOS
Supply voltage	18 V
Die size	6.3 × 4.3 mm²
IO pads	2 × power, 3 × data, 12 × electrodes
Channels	4 tripole
Stimulation	
- Stimulus current amplitude	1 mA, 4 mA or 8 mA
- Anode current ratio adjustment	±20% in steps of 2%
- Stimulus pulse duration	2 µs to 1.07 ms
Temperature sensor resolution	0.12 °C
Humidity sensor resolution	~1%RH
Voltage sensor	
- Detectable range	0 – 18 V
- Resolution	9 steps

1. Temp sensor
2. Electrode voltage / supply voltage sensor
3. Capacitive humidity sensor
4. Digital Circuits
5. Seal ring for micropackaging
6. ESD protection for I/Os

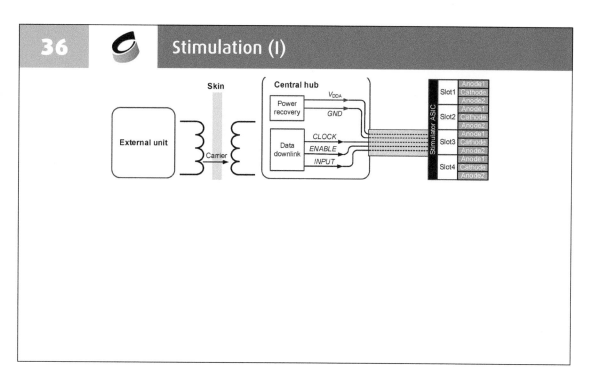

36   Stimulation (I)

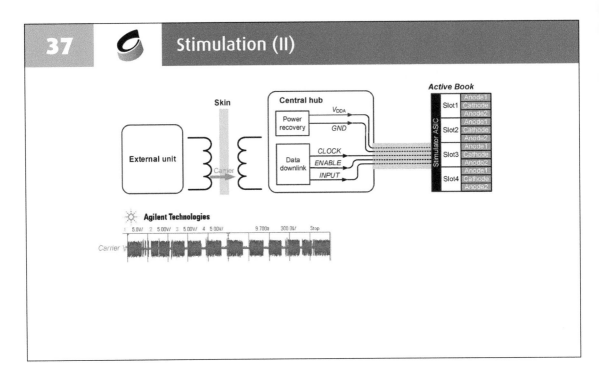

37 Stimulation (II)

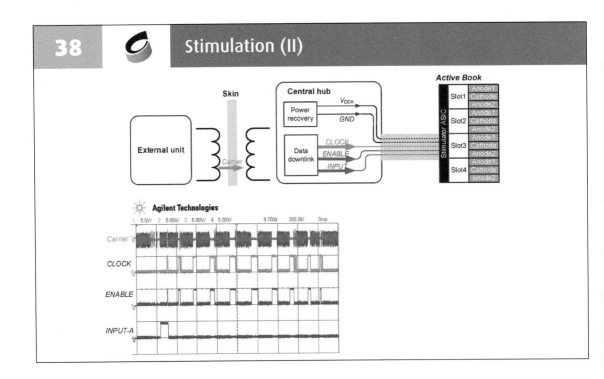

38 Stimulation (II)

39 Stimulation (III)

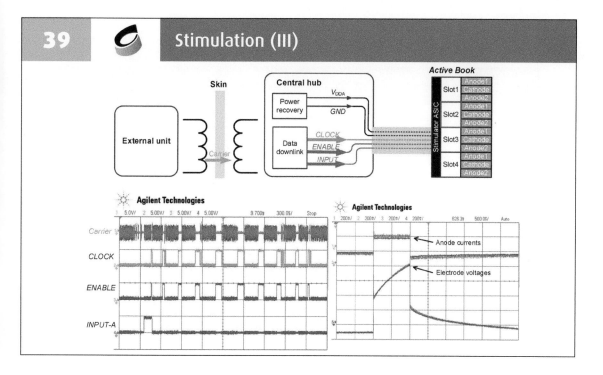

40 Humidity Sensing

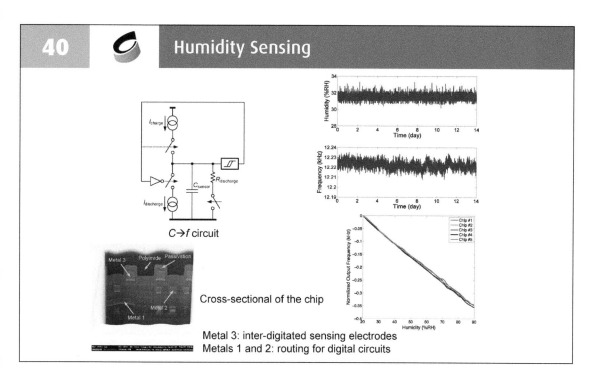

$C{\rightarrow}f$ circuit

Cross-sectional of the chip

Metal 3: inter-digitated sensing electrodes
Metals 1 and 2: routing for digital circuits

41 Comparison of Scalable Stimulation System using Networked ASICs (I)

Ref	Application	Network Topology	Stimulation DeMUX Type	ASICs per System	Channels per System	Inputs to Each ASIC	Methods for Addressing ASICs and Channels
[1] [2]	Cortical Stimulation	Daisy chain	Regenerative	32	2048 (64 per ASIC)	2	ASICs cannot be individually addressed. By channel ID
[3]	Spinal cord Stimulation	Daisy chain	Direct	3	12 (4 per ASIC)	3	By ASIC ID and channel ID
[4]	Retinal Stimulation	Daisy chain	Regenerative	9	9 (1 per ASIC)	4	By ASIC ID and dedicated channel inputs
[5]	Brain Implant	Daisy chain	Regenerative	Up to 63	(6 stim + 4 rec per ASIC)	4	By ASIC ID
[6]	Vestibular Stimulation	Star-shape	Regenerative	3	18 (6 per ASIC)	5	By dedicated ASIC inputs and channel ID
[7] [8]	Spinal cord Stimulation	Star-shape	Regenerative	4	16 (4 per ASIC)	5	By dedicated ASIC inputs and sequential channel addressing
[9]	Cortical Recording	Daisy chain	N/A	2	64 (32 per ASIC)	6	By ASIC ID and channel ID
[10]	Spinal cord Stimulation	Star-shape for addressing. Daisy chain for stimulus current	Direct	9 + 1	27 (3-6 per ASIC)	8	By dedicated multiplexer inputs and channel ID
[11]	Functional Electrical stimulation	Daisy chain	Regenerative	16	512 (32 per ASIC)	9	By sequential ASIC addressing and channel ID

42 Comparison of Scalable Stimulation System using Networked ASICs (II)

In the network topology, the daisy chain is a wiring scheme in which multiple ASICs are connected together in sequence or in a ring. All ASICs share the same input in a daisy chain. In a star-shape configuration, the input to individual ASICs from a central device has some wires common to all ASICs and some ASIC-specific wires.

References
[1] Ghovanloo, *IEEE J. Solid-State Circuits*, 39(12), pp. 2457-2466, 2004
[2] Ghovanloo, *IEEE Trans. Neural Syst. Rehabil. Eng.*, 15(3), pp. 449-457, 2007
[3] Giagka, *IEEE Trans. Biomed. Circuits Syst.*, 9(3), pp. 387-400, 2015
[4] Noda, *Electron. Lett.*, 48(21), pp. 1328-1329, 2012
[5] Ghoreishizadeh, *IEEE Trans. Circuits Syst. I, Reg. Papers*, 64(12), pp. 3056-3067, 2017
[6] Jiang, *IEEE Trans. Biomed. Circuits Syst.*, 9(1), pp. 124-137, 2015.
[7] Liu, *IEEE Trans. Biomed. Circuits and Syst.*, 6(3), pp. 216-227, 2012.
[8] Liu, *IEEE Design & Test*, 33(4), pp. 8-23, 2016
[9] Sodagar, *IEEE J. Solid-State Circuits*, 44(9), pp. 2591-2604, 2009.
[10] Gad, *J. Neuroeng. Rehabil*, 10(2), pp. 1-17, 2013.
[11] Meza-Cuevas, *Proc. 2014 Middle East Conf. Biomed. Eng. (MECBME)*, pp. 111-114, 2014.

43 Conclusion

Future implant system:

1. A network of small implants
2. Individual implants will be smaller and offer finer interaction with neural environment
3. Reliable and safe

Thank you.

xiao@ucl.ac.uk

CHAPTER 07

CAS for Control of Prosthetic Hands

Kia Nazarpour

Newcastle University

In this talk we will review briefly CAS-related developments in the area of prosthetic limb and share examples of success, in measurement of neuro-muscular activity, real-time processing and sensory feedback. We conclude this talk by discussing challenges and future trends in prosthetics control.

1   Upper-limb loss statistics

- In the UK
  - 473 new upper-limb referrals every year
  - 245 in the age range of 15-54
- In the US
  - overall 500,000 people with limb loss
- Every 2,500 people born with upper-limb reduction

2   Cosmetic Hands and digits

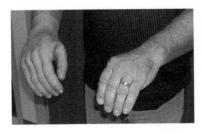

Before   Silicone Fingers   After

Before   Test Piece

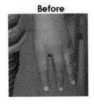

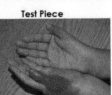

Courtesy of Sarah Day

3 **History (I)**

Iron Prosthetic Hand **1560-1600**
Wellcome Library, London

Ambroise Paré
(c. 1510 – 1590)
French barber surgeon

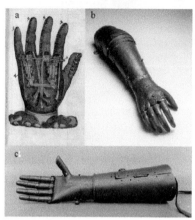

Iron Prosthetic Hand (owned by a German knight) **c. 1504**

---

4 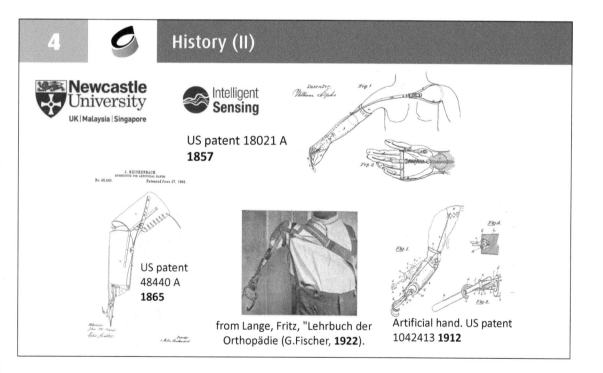 **History (II)**

Newcastle University
UK | Malaysia | Singapore

Intelligent Sensing

US patent 18021 A
1857

US patent
48440 A
1865

from Lange, Fritz, "Lehrbuch der Orthopädie (G.Fischer, **1922**).

Artificial hand. US patent
1042413 **1912**

136

5 Early Bionic Hands (I)

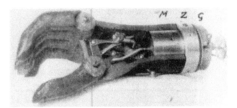

Electric powered hand **1948**
Germany

SVEN-Hand **1965**
Sweden

6 Early Bionic Hands (II)

Southampton Hand **1990's**
UK

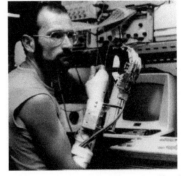

First Microprocessor-controlled hand
1990's, UK

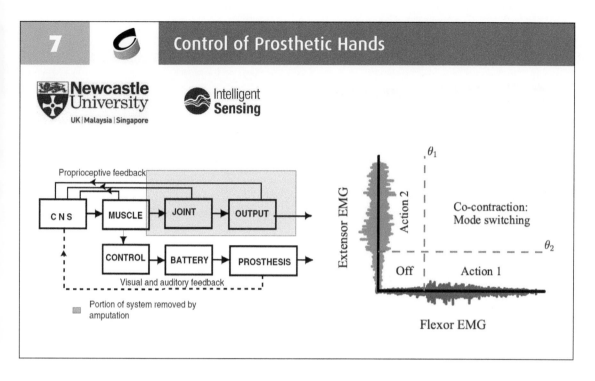

7 — Control of Prosthetic Hands

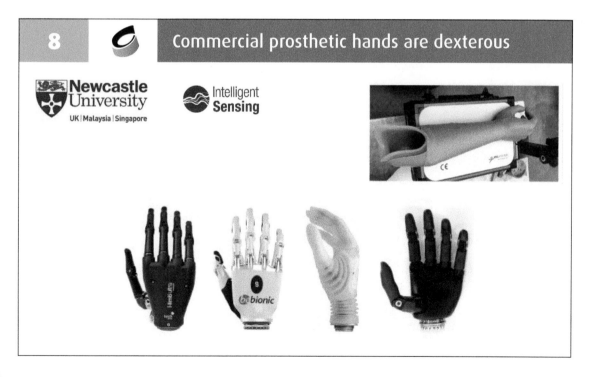

8 — Commercial prosthetic hands are dexterous

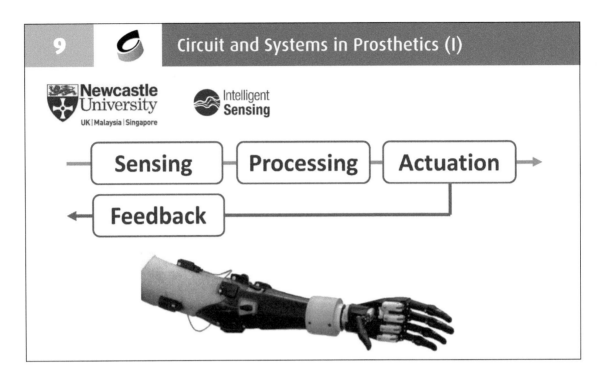

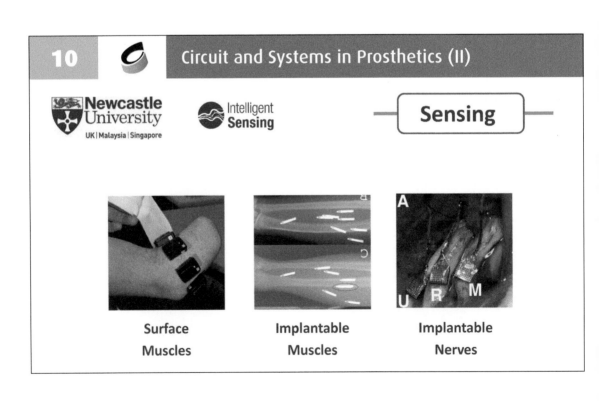

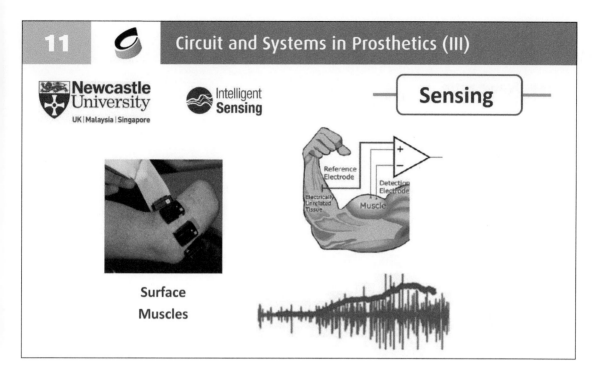

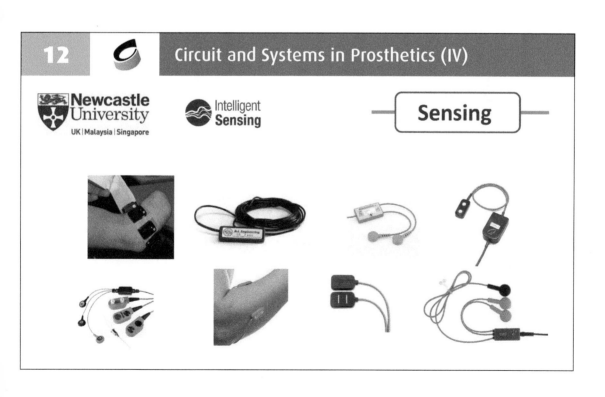

13 Circuit and Systems in Prosthetics (V)

Newcastle University
UK | Malaysia | Singapore

Intelligent **Sensing**

Sensing

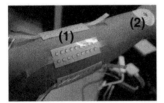
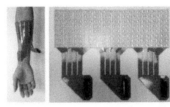

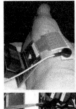

14 Circuit and Systems in Prosthetics (VI)

Newcastle University
UK | Malaysia | Singapore

Intelligent **Sensing**

Sensing

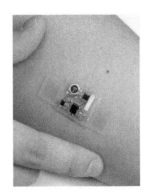

15   Circuit and Systems in Prosthetics (VII)

 Intelligent Sensing

Sensing

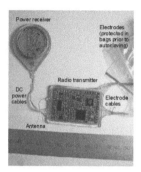

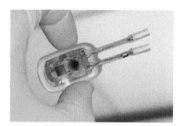

16   Circuit and Systems in Prosthetics (VIII)

 Intelligent Sensing

Sensing

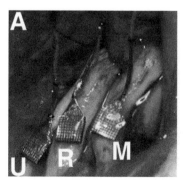

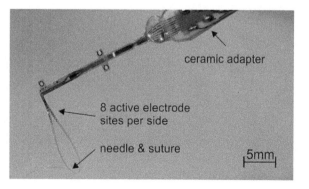

17 Where are we?

 Sensing

- We have almost reached excellent control levels from the surface of the skin

MEC 2017 Award

- Krasoulis et al, JNER 2017

18 Challenges

 Sensing

- In order to increase dexterity control we need [reliable] implantable, wireless technology

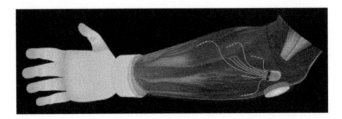

19 **Circuit and Systems in Prosthetics (I)**

 **Processing**

- Innovation – very little!

- Typically involves, pre-processing and signal conditioning

- Pattern Recognition
  - Feature Extraction
  - Classification

20 **Circuit and Systems in Prosthetics (II)**

 **Processing**

- Innovation – almost NONE!

- Vast, open field

- Ideas desperately needed

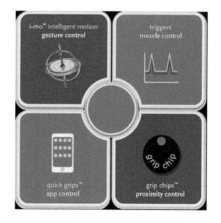

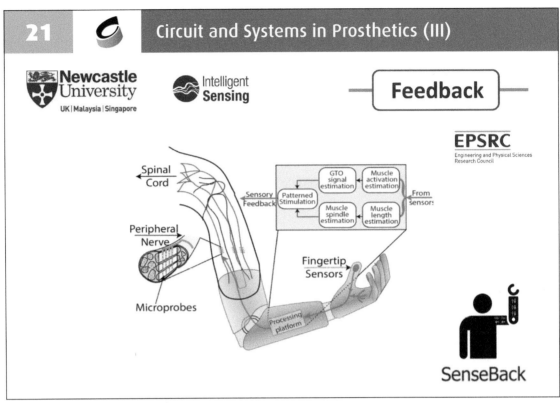

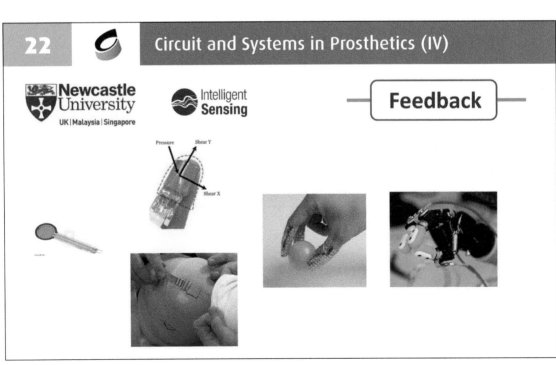

23 Circuit and Systems in Prosthetics (V)

Feedback

Non-invasive

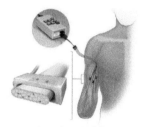

Implantable Nerves

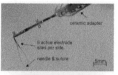

Implantable Nerves

---

24 Challenges

Feedback

- Innovation
  - significant
- Impact
  - very little
- Studied, especially the invasive ones, are performed within hospital environment, with little control and hence limited reproducibility

25 Future Trends

Newcastle University
UK | Malaysia | Singapore

Intelligent Sensing

Sensing — Processing — Actuation
Feedback

- For biomedical engineers, it is not anymore enough to only be engineers.
  - Science, Biology, Experiment Design

- Stake-holders
  - Patient, Clinicians, Industry

- Evidence

26 Prosthetic Control –Current trends

Newcastle University
UK | Malaysia | Singapore

Intelligent Sensing

- **An engineering approach**

- **An alternative approach**

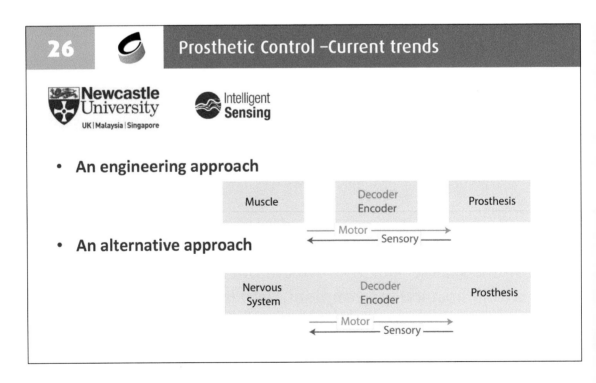

27 Abstract decoding

28 Challenges

✉ **K.Nazarpour@newcastle.ac.uk**

🐦 **KiaNazarpour**

〜 **www.intellsensing.com**

CHAPTER 08

Genetically Enhanced Brain-implants for Neuro-rehabilitation

Patrick Degenaar

Newcastle University

Neuroprosthetics have been steadily evolving since the early cardiac pacemaker development in the late 1950's. The late 1970's saw the emergence of cochlear implants and the 1990's saw the emergence of deep brain pacemakers for motor disorders. Such technologies have been based on electrical recording and stimulus of the nervous system to provide therapeutic benefit. The key challenges centre on providing the correct form of stimulus and ensuring that the devices do not corrode to dysfunction inside the body.

In 2013, demonstrations of channel rhodopsin and melanopsin genetically inserted into nerve cells were presented in the literature. These discoveries, which led to the field of optogenetics, allowed nerve cells to be controlled by light, opening up new ways of communication. It typically takes two decades for novel discoveries to get to clinical practice. The field of optogenetics is now 15 years old, and trials in humans have already begun in the human retina. The challenge is how to achieve a much more difficult task of developing optogenetic neuroprosthetics which can be used for visual brain prosthetics and treat conditions such as epilepsy.

But with new opportunities come new challenges. In tandem with developments of neurotechnology, the regulatory and ethical requirements have also increased. Demonstration of safety and efficacy is paramount. As such, the first demonstration for this technology was in the retina of people blinded by Retinitis Pigmentosa. i.e. optical technology could be external to the eye, and should anything go wrong, removal of the eye causes no harm to the patient. To progress towards brain disorders, the primary initial target condition is epilepsy, as current clinical practice for drug-resistant focal epilepsy is to remove the problematic part of the brain provided it is non-eloquent. It is, therefore, the focus of the CANDO project which aims to achieve clinical trials in the early 2020's.

This talk provides a full discussion of the opportunities in the field of optogenetics, how it can provide benefit over existing electrical neural prosthesis, and the new challenges in bringing such devices to clinical trials.

1 The science fiction...

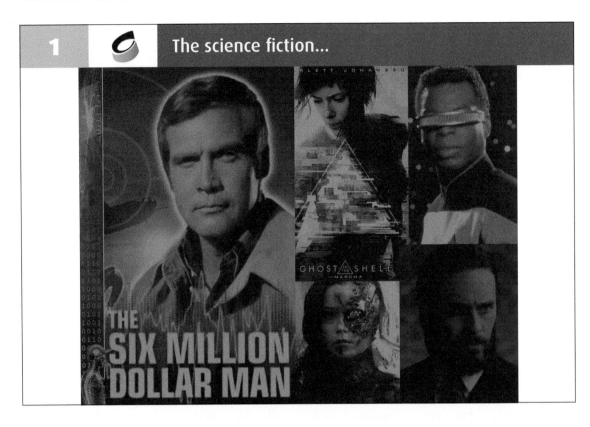

2 Real Neuroprosthetics:
Parkinson's disease exemplar

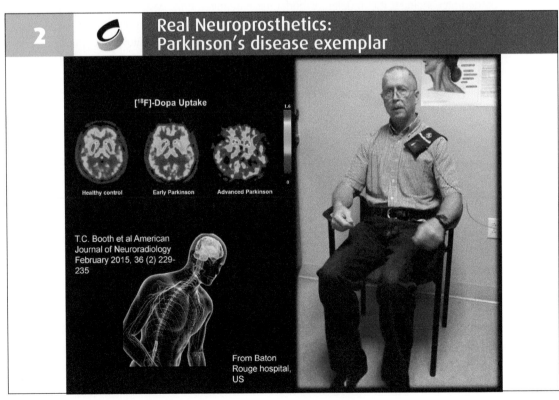

3 — Neuroprosthetic applications

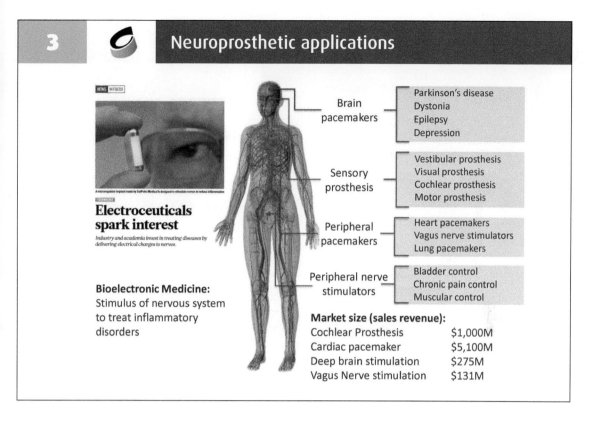

Electroceuticals spark interest

Industry and academia invest in treating diseases by delivering electrical charges to nerves.

Bioelectronic Medicine:
Stimulus of nervous system to treat inflammatory disorders

Brain pacemakers
- Parkinson's disease
- Dystonia
- Epilepsy
- Depression

Sensory prosthesis
- Vestibular prosthesis
- Visual prosthesis
- Cochlear prosthesis
- Motor prosthesis

Peripheral pacemakers
- Heart pacemakers
- Vagus nerve stimulators
- Lung pacemakers

Peripheral nerve stimulators
- Bladder control
- Chronic pain control
- Muscular control

Market size (sales revenue):

Cochlear Prosthesis	$1,000M
Cardiac pacemaker	$5,100M
Deep brain stimulation	$275M
Vagus Nerve stimulation	$131M

4 — Current protocol: Shouting not listening!

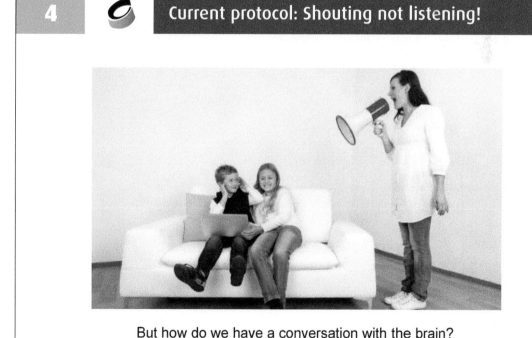

But how do we have a conversation with the brain?

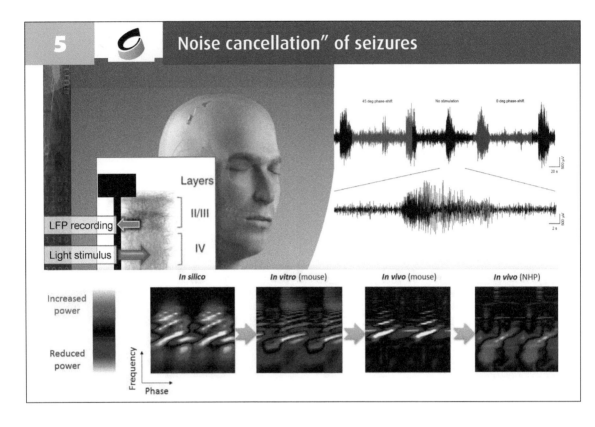

5 Noise cancellation" of seizures

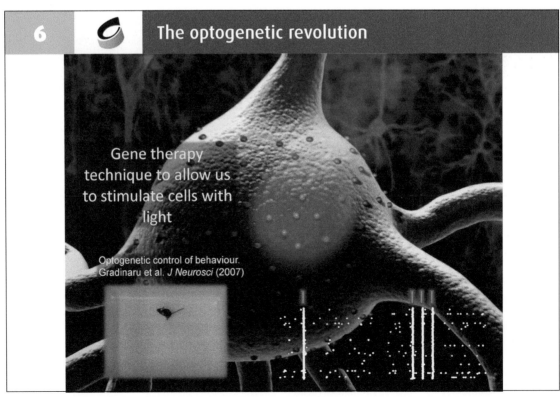

6 The optogenetic revolution

Gene therapy technique to allow us to stimulate cells with light

Optogenetic control of behaviour.
Gradinaru et al. *J Neurosci* (2007)

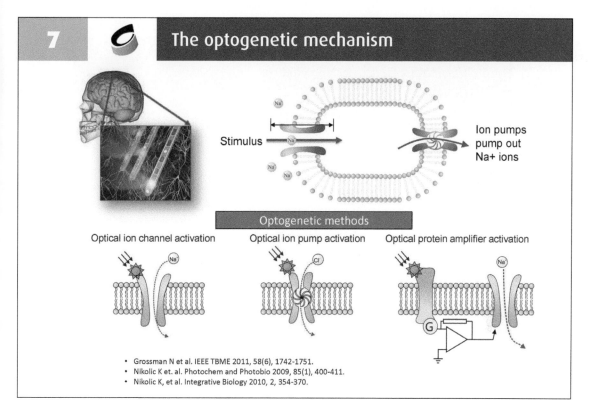

7 The optogenetic mechanism

Stimulus

Ion pumps pump out Na+ ions

Optogenetic methods

Optical ion channel activation   Optical ion pump activation   Optical protein amplifier activation

- Grossman N et al. IEEE TBME 2011, 58(6), 1742-1751.
- Nikolic K et. al. Photochem and Photobio 2009, 85(1), 400-411.
- Nikolic K, et al. Integrative Biology 2010, 2, 354-370.

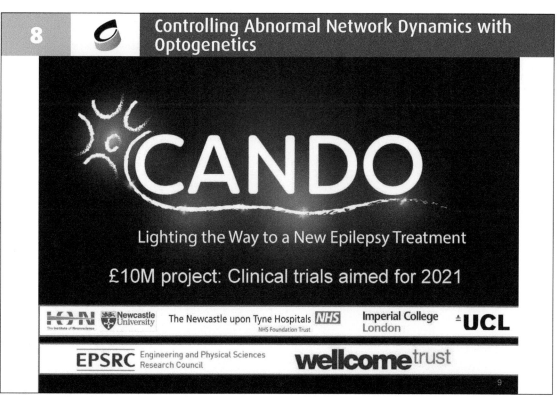

8 Controlling Abnormal Network Dynamics with Optogenetics

CANDO

Lighting the Way to a New Epilepsy Treatment

£10M project: Clinical trials aimed for 2021

Newcastle University   The Newcastle upon Tyne Hospitals NHS   Imperial College London   UCL

EPSRC Engineering and Physical Sciences Research Council   wellcometrust

9

9

Noise cancellation of epileptic seizures – but adaptable to vision

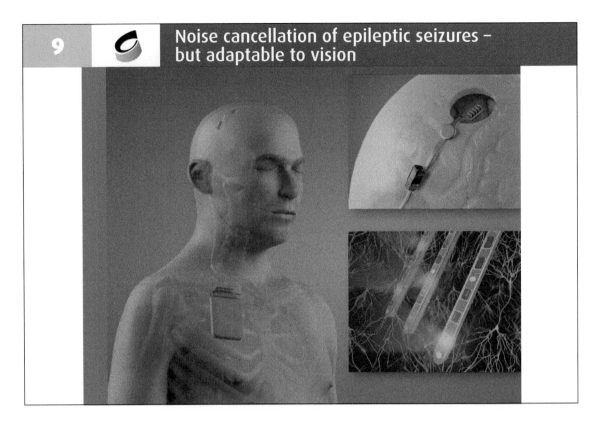

10

Closed loop control of epileptic seizures

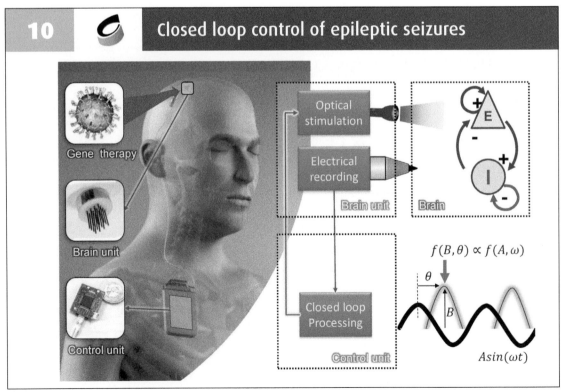

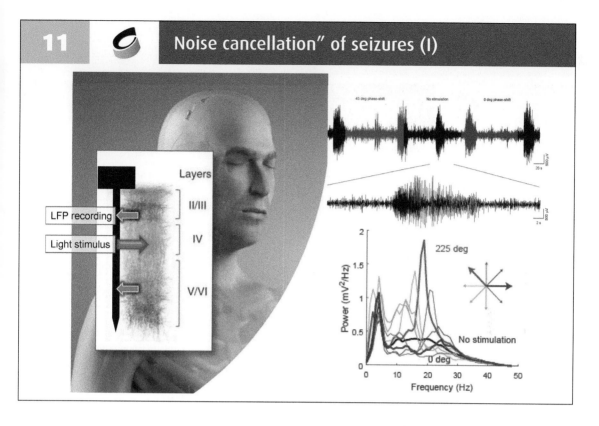

11 Noise cancellation" of seizures (I)

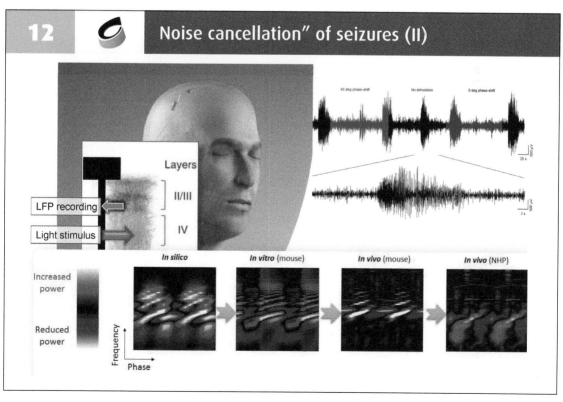

12 Noise cancellation" of seizures (II)

WHAT ARE THE CIRCUIT/SYSTEM CHALLENGES FOR THE CAS COMMUNITY?

13 Light transmission through tissue

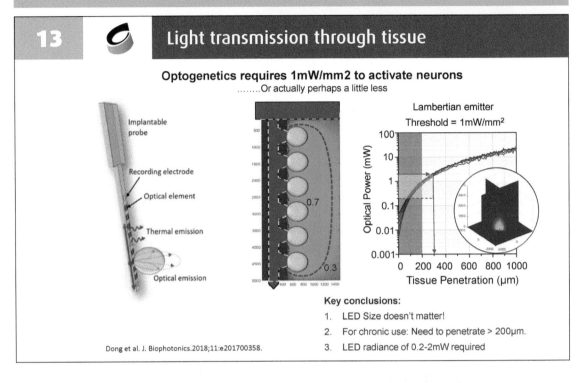

Optogenetics requires 1mW/mm2 to activate neurons
........Or actually perhaps a little less

Implantable probe

Recording electrode

Optical element

Thermal emission

Optical emission

Lambertian emitter
Threshold = 1mW/mm²

Dong et al. J. Biophotonics.2018;11:e201700358.

Key conclusions:
1. LED Size doesn't matter!
2. For chronic use: Need to penetrate > 200µm.
3. LED radiance of 0.2-2mW required

14 LED Efficiency

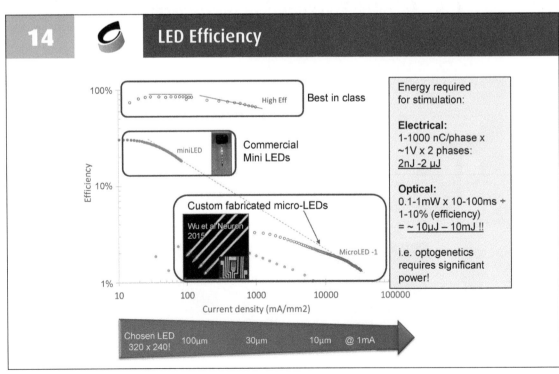

Best in class

Commercial Mini LEDs

miniLED

Custom fabricated micro-LEDs

Wu et al Neuron 2015

MicroLED -1

Chosen LED 320 x 240! 100µm   30µm   10µm   @ 1mA

Energy required for stimulation:

Electrical:
1-1000 nC/phase x ~1V x 2 phases:
2nJ -2 µJ

Optical:
0.1-1mW x 10-100ms ÷ 1-10% (efficiency)
= ~ 10µJ – 10mJ !!

i.e. optogenetics requires significant power!

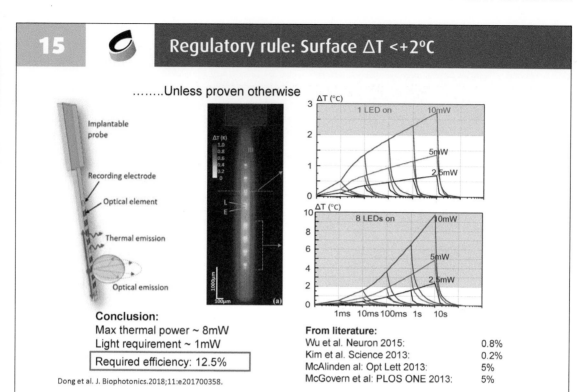

15   **Regulatory rule: Surface ΔT <+2°C**

........Unless proven otherwise

Conclusion:

Max thermal power ~ 8mW

Light requirement ~ 1mW

Required efficiency: 12.5%

Dong et al. J. Biophotonics.2018;11:e201700358.

From literature:

Wu et al. Neuron 2015:	0.8%
Kim et al. Science 2013:	0.2%
McAlinden al: Opt Lett 2013:	5%
McGovern et al: PLOS ONE 2013:	5%

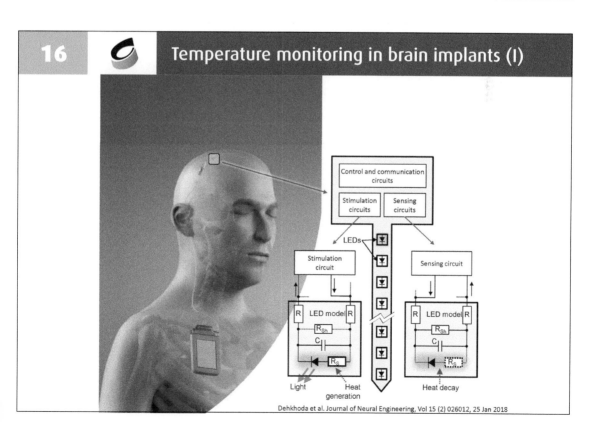

16   **Temperature monitoring in brain implants (I)**

Dehkhoda et al. Journal of Neural Engineering, Vol 15 (2) 026012, 25 Jan 2018

17   Temperature monitoring in brain implants (II)

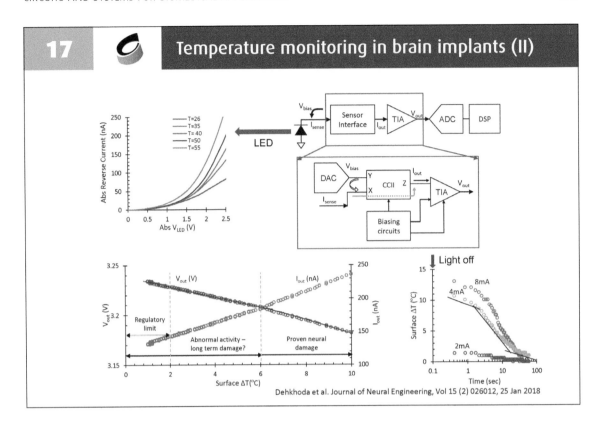

Dehkhoda et al. Journal of Neural Engineering, Vol 15 (2) 026012, 25 Jan 2018

18   Intelligent brain stimulator concept (I)

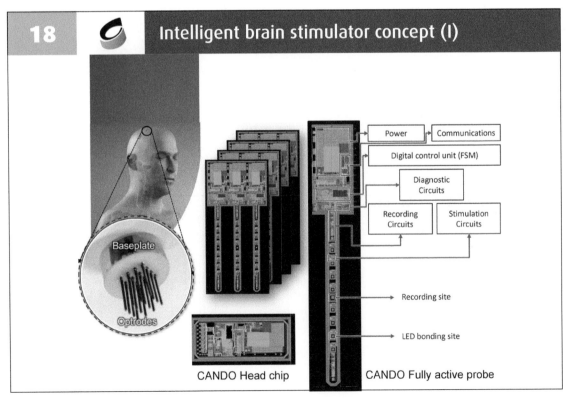

CANDO Head chip

CANDO Fully active probe

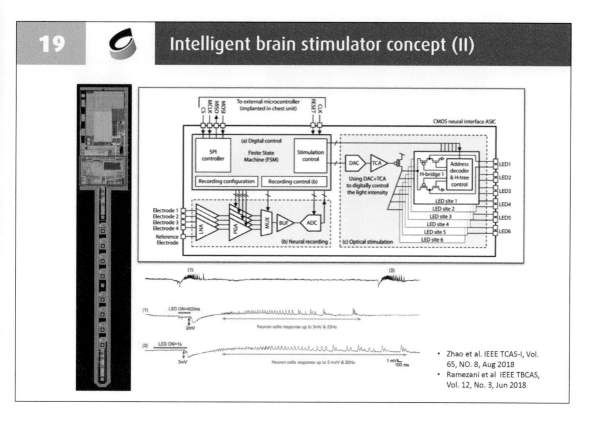

19 — Intelligent brain stimulator concept (II)

- Zhao et al. IEEE TCAS-I, Vol. 65, NO. 8, Aug 2018
- Ramezani et al IEEE TBCAS, Vol. 12, No. 3, Jun 2018

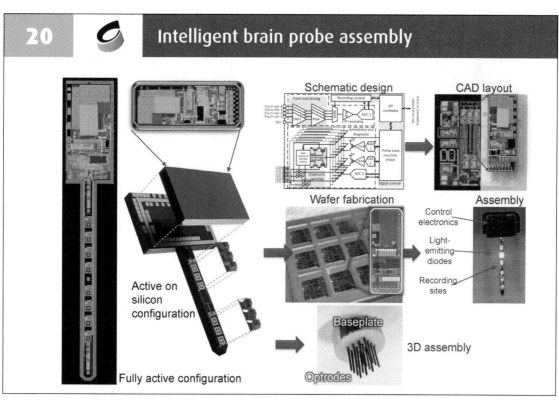

20 — Intelligent brain probe assembly

Schematic design

CAD layout

Wafer fabrication

Assembly
- Control electronics
- Light-emitting diodes
- Recording sites

Active on silicon configuration

Fully active configuration

Baseplate

Optrodes

3D assembly

21 Intelligent brain probe assembly

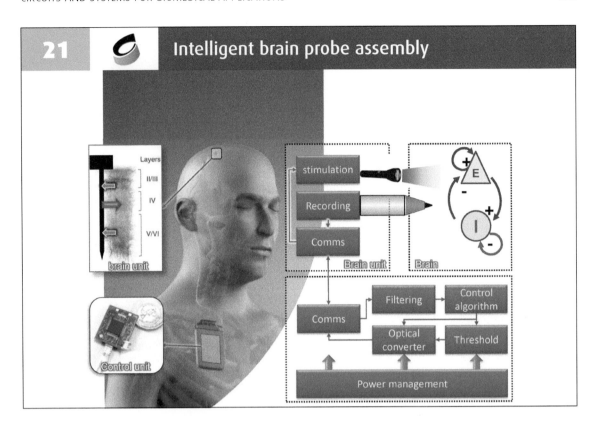

22 CANDO Closed loop control system

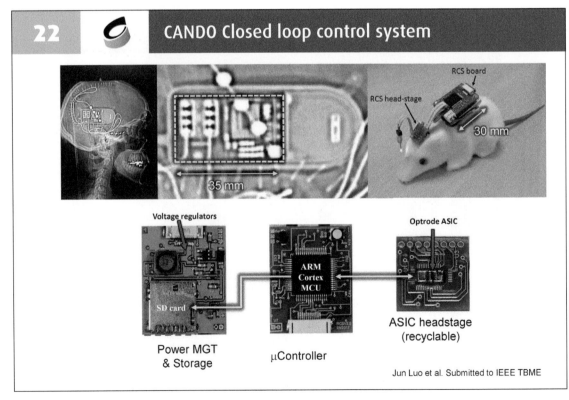

Jun Luo et al. Submitted to IEEE TBME

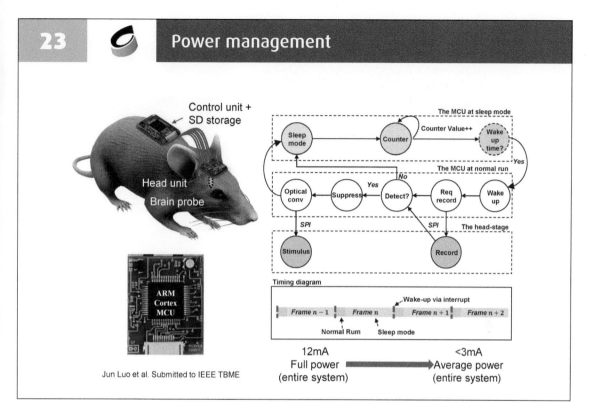

23 Power management

Control unit + SD storage

Head unit

Brain probe

ARM Cortex MCU

Jun Luo et al. Submitted to IEEE TBME

Timing diagram

12mA
Full power
(entire system)

<3mA
Average power
(entire system)

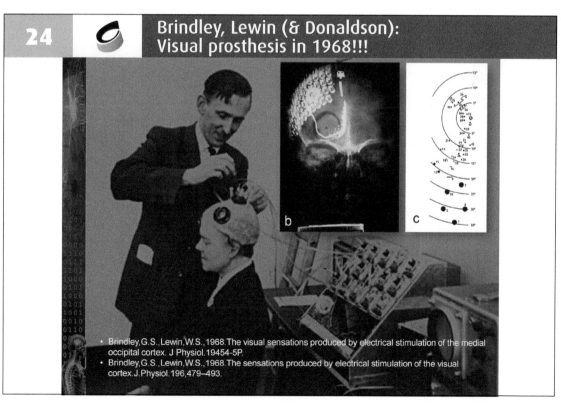

24 Brindley, Lewin (& Donaldson): Visual prosthesis in 1968!!!

- Brindley,G.S.,Lewin,W.S.,1968.The visual sensations produced by electrical stimulation of the medial occipital cortex. J Physiol.19454-5P.
- Brindley,G.S.,Lewin,W.S.,1968.The sensations produced by electrical stimulation of the visual cortex.J.Physiol.196,479–493.

25 Optogenetic visual cortical prosthetics (I)

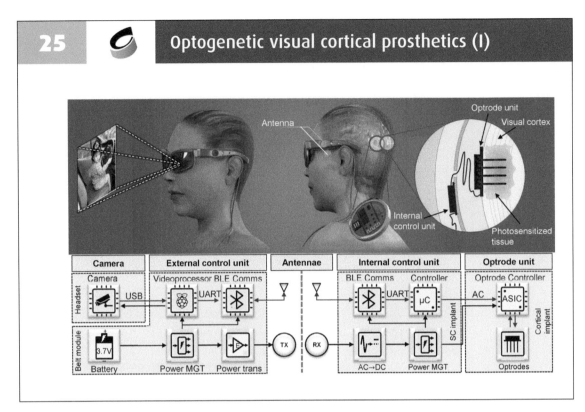

26 Optogenetic visual cortical prosthetics (II)

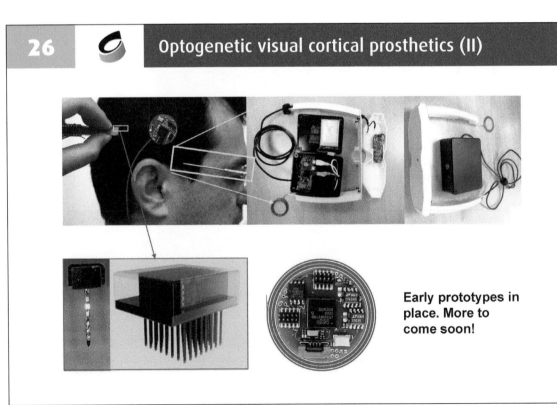

Early prototypes in place. More to come soon!

27 Acknowledgements

Team: Oct 2018

About
the Editors

Hadi Heidari

Dr Hadi Heidari is a Lecturer (Assistant Professor) in the School of Engineering and lead of the Microelectronics Lab (meLAB) at the University of Glasgow. He received his PhD in Microelectronics from the University of Pavia (Italy) in 2015, where he worked on Integrated CMOS Magnetic Sensory Microsystems. He spent Postdoctoral at the University of Glasgow before he joined the Glasgow College UESTC in 2016.

Dr Heidari is member of the IEEE Circuits and Systems Society Board of Governors (BoG), IEEE Sensors Council Administrative Committee (AdCom), IEEE Sensors Council Young Professional Representative and Senior Member of IEEE. He is on the Editorial Board of Microelectronics Journal, Guest Editor for the IEEE Sensors Journal, and Guest Associate Editor for the IEEE Journal of Electromagnetics, RF and Microwaves in Medicine and Biology and IEEE Access. He serves on the organising committee of several conferences including the UK-China Emerging Technologies (UCET) Conference, IEEE SENSORS'16 and '17, NGCAS'17, BioCAS'18, PRIME'15 and '19, and the organiser of several special sessions on the IEEE Conferences. His research has been funded by major research councils and funding organizations including the European Commission, EPSRC, Royal Society and Scottish Funding Council. He is part of the €8.4M EU H2020 FET Proactive project on "Hybrid Enhanced Regenerative Medicine Systems (HERMES)".

Dr Heidari has authored/co-authored over 80 peer-reviewed publications in international journals or conference proceedings and acts as a reviewer for several journals and conferences. He has been the recipient of a number of awards including the IEEE CASS Scholarship (NGCAS'17 conference), Silk Road Award from the Solid-State Circuits Conference (ISSCC'16), Best Paper Award from the IEEE ISCAS'14 conference, Gold Leaf Award from the IEEE PRIME'14 Conference and Rewards for Excellence prize from UofG (2018). He was a research visitor with the University of Macau, China, and McGill University, Canada.

hadi.heidari@glasgow.ac.uk

Sara Ghoreishizadeh

Dr Sara Ghoreishizadeh is a lecturer in wearable technology at University College London since November 2018. She received a PhD from EPFL, Switzerland in 2015 and has been with the Dept. of Electrical and Electronic Engineering, Imperial College as a Junior Research Fellow until Oct 2018. She received her BSc and MSc degrees (both with distinction) in Microelectronics from Sharif University of technology in 2007 and 2009, respectively. Her research interests include analogue/mixed-signal circuits design for wearable and implantable biosensors and the monolithic integration of biosensor with microelectronics, on which she has more than 30 peer-reviewed publications in top IEEE journal and conference proceedings and supervised many(MEng, MSc, PhD) students. She has received the EPSRC eFutures Early Career Researcher award in 2018 to lead a multidisciplinary team on creating an autonomous wearable biosensor device. She has been on the technical programme committee and review committee of IEEE BioCAS 2017-2018 and IEEE ICECS 2016-2018 conferences. She has been an associate editor of the Journal of Microelectronics since 2015, and an elected member of IEEE BioCAS technical committee since 2018. She is a (co-)organiser of the first UKCAS in 2018.

s.ghoreishizadeh@ucl.ac.uk

About
the Authors

David Cumming

Prof. David Cumming is Head of the School of Engineering and the Professor of Electronic Systems at the University of Glasgow. He completed his B.Eng in EEE at Glasgow University in 1989 and his Ph.D. in microelectronics at Cambridge University in 1993. He subsequently worked as a VLSI design engineer with STMicroelectronics in Bristol, UK, and carried out post-doctoral research at Glasgow University. He was a lecturer at the University of Canterbury, NZ, and has been at Glasgow University since 1999. His research interests are focused on semiconductor devices and microsystems for sensing applications. He has published over 300 papers on sensor and photonic technologies, mainly with an emphasis on integration of technologies. He was a founder of medical diagnostics company Mode DX and his research on CMOS ion sensor arrays enabled the development of the Ion Torrent gene sequencing system. He has held 1851 and EPSRC research fellowships, a Chinese Academy of Science Distinguished Fellowship, a Royal Society Wolfson Merit Award, and is FRSE, FREng, FIEEE. He has served on the technical programme committees for the EIPBN, MNE and IEEE IWASI conferences, and will be the Plenary Session Chair for ISCAS 2023 in Monterey California.

[Chapter 1]

Patrick Degenaar

D r Patrick Degenaar is a reader in biomedical engineering and came to Newcastle in 2010 to develop world class collaborations between Electrical and Electronic Engineering at Newcastle University and the Institute of Neuroscience. He has a BSc (1st class) and MRes in Applied Physics from Liverpool University, and a PhD in Bioimaging from the Japan Advanced Institute for Science and Technology. After some time in the software industry, he did two post-doctoral projects at Imperial College before getting an RCUK fellowship in 2005. From 2005-2010 he was a lecturer and then senior lecturer at Imperial College, before coming to Newcastle. He has had numerous research awards and published numerous papers in the key journals in the biomedical field.

At the heart of these efforts is my pioneering use of CMOS-micro-LED optoelectronics in combination with optogenetic gene therapy solutions. These will lead to highly advanced forms of prosthetic intervention not previously possible. This has led to a number of highly cited papers in key biomedical engineering journals. Furthermore, I have explored impact through patient trials and commercial translation.

He has been part of a number of large research consortia. Between 2010-2014 he coordinated the FP7 OptoNeuro project. More recently he is the engineering team leader on the £10M CANDO project to develop a next-generation prosthesis for epilepsy. Currently he has a large highly dedicated team of RAs, and PhD students.

[Chapter 8]

Marc Desmulliez

Prof. Marc Desmulliez, FRSE, FIET, FInstP, Ph.D., is an electrical and electronic engineer by background and holds also 2 Master of Science degrees in Microwave and Modern Optics (UCL) and Theoretical Physics (Cambridge). He is currently Professor in Microsystems Engineering at Heriot-Watt University, where he leads the Multimodal Sensing and Micro-Manipulation (CAPTURE) Research Group. He is the Director of the Nature Inspired Manufacturing Centre at Heriot-Watt University. Co-authors of around 460 publications, his current research interests span medical device technology, novel additive manufacturing processes, microwave sensing and biomimetics.

He span out or started up 3 Companies, MicroStencil (January 2003), MicroSense Technologies Limited (January 2017) and Birthing Solutions Ltd (November 2017). The first company specialised in the manufacturing of electroformed stencils for the microelectronics industry. The second company manufactures microwave sensors for the food & drink industries and shared the £90K first prize of the Converge Challenge 2016, the most prestigious prize in Entrepreneurship in Scotland.

Prof. Desmulliez also created the first UK Master of Science Microsystems Engineering in 2001 which still continues today as the Erasmus Mundus MSc in Smart Systems Integration.

[Chapter 2]

Steve Hall

Prof Steve Hall was Head of the Department of Electrical Engineering and Electronics at the University of Liverpool from 2001 to 2009 and is currently the Director of Research. He has interests spanning materials characterization, device physics & innovative device design and gate level circuits. He has about 300 conference and journal papers in these areas including novel measurements and contributions to the understanding of MOS related interfaces and materials quality. He successfully designed and built novel MOS and bipolar devices in silicon for about 20 years. More recently, his work encompasses high permittivity dielectrics, conducting oxides, rectennas for energy scavenging and biologically inspired device/circuit concepts. Prof. Hall has served on the Steering and Programme Committees of ESSDERC/ESSCIRC and INFOS and involved with the organization of both. He is an associate Editor of IEEE Electron Device Letters.

[Chapter 4]

Xiao Liu

Dr Xiao Liu received the B. Eng. degree in Information Engineering from Xi'an Jiaotong University, China, in 2003, the M.Sc. degree in Microelectronics Systems Design from the University of Southampton, U.K., in 2004, and the Ph.D. degree from University College London (UCL), U.K., in 2008. From 2009 to 2011, he was a Research Associate with the Analogue and Biomedical Electronics Group, UCL. From 2011 to 2013, he was a lecturer with the School of Engineering and Design, Brunel University, U.K. He moved back to UCL in 2013 and is currently a Lecturer with the Department of Electronic and Electrical Engineering.

He is an Associate Editor of the IEEE Transactions on Circuits and Systems I: Regular Papers. He is a Chartered Engineer and a member of the Biomedical and Life Science Circuits and Systems Technical Committee of the IEEE Circuits and Systems Society.

[Chapter 6]

Liam J. McDaid

Liam J. McDaid received his B.Eng. (Hons) degree in electrical and electronics engineering from the University of Liverpool in 1985 and in 1989 he received a PhD in solid-state devices from the same institution. He is currently Professor of Computational Neuroscience at Ulster University and leads the Computational Neuroscience and Neural Engineering (CNET) Research Team. His current research interests include modelling the role of glial cells in the functional and dysfunctional brain. He is also involved in the development of software/hardware models of neural-based computational systems, with particular emphasis on the mechanisms that underpin self-repair in the human brain. He has received several research grants in this domain and is currently a collaborator on an EPSRC funded project. He has co-authored over 120 publications in his career to date.

[Chapter 4]

Srinjoy Mitra

Srinjoy Mitra is a Senior Lecturer at the University of Edinburgh, UK. He received master's degree in electronics from the Indian Institute of Technology, Bombay, India, in 2003. After briefly working in microelectronic industry, he completed his Ph.D from the Institute of Neuroinformatics (UNI), ETH Zurich, Switzerland, in 2008. Between 2008 and 2010, he worked as a Postdoctoral Researcher at Johns Hopkins University, Baltimore, MD, USA. He then joined the medical electronics team at imec, Belgium, and worked there as a Senior Scientist until early 2016. Dr. Mitra has worked in neuromorphic electronics and low-power ICs for prosthetic applications. At imec he had taken up lead roles in a number of industrial and publicly funded projects primarily related to biopotential recording. Various ICs developed during these projects have been licensed to industry. For the last few years Dr. Mitra led multiple projects on neural implants for central and peripheral nervous system. His primary research interest is in designing novel mixed signal CMOS circuits for advancement in medical and neural electronics.

[Chapter 5]

Kianoush Nazarpour

Dr Kianoush Nazarpour received the B.Sc. degree from the K. N. Toosi University of Technology, Tehran, Iran, in 2003, the M.Sc. degree from Tarbiat Modarres University, Tehran, in 2005, and the Ph.D. degree from Cardiff University, Cardiff, U.K., in 2008, all in electrical and electronic engineering. From 2007 to 2012, he held two post-doctoral researcher posts at the University of Birmingham and Newcastle University. In 2012, he joined Touch Bionics Inc., U.K., as a Senior Algorithm Engineer working on intelligent control of multifunctional myoelectric prostheses. In 2013, he returned to Newcastle University, where he is currently a Reader in biomedical engineering. His research interests include intelligent sensing and biomedical signal processing and their applications in assistive technology. He received the Best Paper Award at the 3rd International Brain-Computer Interface Conference, Graz, Austria (2006), and the David Douglas Award (2006), U.K., for his work on the joint space–time–frequency analysis of the electroencephalogram signals. He is currently an Associate Editor of the Medical Engineering and Physics Journal in the area of biomedical signal processing.

[Chapter 7]

Themis Podromakis

Themis Prodromakis is Professor of Nanotechnology and Head of the Electronic Materials and Devices Research Group in the Zepler Institute, University of Southampton, UK. His work focuses on developing metal-oxide Resistive Random-Access Memory technologies and related applications and is leading an interdisciplinary team comprising 15 researchers with expertise ranging from materials process development to electron devices and circuits and systems for embedded applications. He holds a Royal Society Industry Fellowship and is a Visiting Professor at the Department of Microelectronics and Nanoelectronics at Tsinghua University, CN and Honorary Fellow at Imperial College London. He is Fellow of the IET, Fellow of the Institute of Physics, Senior Member of the IEEE and serves as the Director of the Lloyds Register Foundation International Consortium for Nanotechnology (ICoN: www.lrf-icon. com). In 2015, Prof Prodromakis established ArC Instruments Ltd, a start-up that delivers high-performance testing infrastructure for automating characterisation of novel nanodevices.

[Chapter 3]